# PRAISE FOR TED L. GUNDERSON
## WHO CAN TELL YOU
## <u>HOW TO LOCATE ANYONE, ANYWHERE</u>

"From December 1951 to August 1979, Ted Gunderson's ability was recognized by his superiors in the Federal Bureau of Investigation by appointments to management and leadership offices in the organization. I congratulate him on his fine career."—President Gerald R. Ford

"No one knows more about finding people."
—Efrem Zimbalist, Jr.

"Ted Gunderson is a professional investigator with outstanding abilities and integrity. His book provides help for those who cannot afford a professional and do not know how to locate a lost loved one."—Peter Ueberroth

"Ted Gunderson is a good detail man. . . . He gets the job done."—Mike Ditka

"Ted L. Gunderson's credentials as an investigator are impeccable. . . . His advice will be most helpful to many people."
—Johnny Carson

**TED GUNDERSON** is a former FBI agent who is now a private investigator. **ROGER MCGOVERN** is a freelance writer. They both live in Palm Springs, CA.

# HOW TO LOCATE ANYONE ANYWHERE

## Without Leaving Home

REVISED EDITION

## TED L. GUNDERSON
### with ROGER McGOVERN

A PLUME BOOK

PLUME
Published by the Penguin Group
Penguin Books USA Inc., 375 Hudson Street, New York, New York 10014, U.S.A.
Penguin Books Ltd, 27 Wrights Lane, London W8 5TZ, England
Penguin Books Australia Ltd, Ringwood, Victoria, Australia
Penguin Books Canada Ltd, 10 Alcorn Avenue, Toronto, Ontario, Canada M4V 3B2
Penguin Books (N.Z.) Ltd, 182–190 Wairau Road, Auckland 10, New Zealand

Penguin Books Ltd, Registered Offices:
Harmondsworth, Middlesex, England

Published by Plume, an imprint of Dutton Signet,
a division of Penguin Books USA Inc.
Original edition published by E.P. Dutton

First Plume Printing, June, 1991
First Plume Printing (Revised Edition), November, 1996
10  9  8  7  6  5  4  3  2  1

*Note:* Fictitious names have been given to individuals represented in anecdotes and other material in this book. They do not relate to real owners of those names in any way and such relationship is coincidental.

*Family History Library and Family History Centers* on pages 122–27 and 143–197. Copyright © 1986, 1988 by Corporation of the President of The Church of Jesus Christ of Latter-day Saints. Reprinted by permission. ■ Family History Library Catalog, Family Registry, International Genealogical Index, Ancestral File, and Personal Ancestral File are trademarks of Corporation of the President of The Church of Jesus Christ of Latter-day Saints. ■ Neither the Family History Library, the Family History Department of The Church of Jesus Christ of Latter-day Saints, nor the Genealogical Society of Utah approves or endorses this book or its contents.

℗ REGISTERED TRADEMARK—MARCA REGISTRADA

LIBRARY OF CONGRESS CATALOGING-IN-PUBLICATION DATA
Gunderson, Ted L.
    How to locate anyone anywhere without leaving home / Ted L. Gunderson with Roger McGovern.—Rev. ed.
        p.  cm.
    Originally published: New York : Dutton, 1989.
    ISBN 0-452-27742-6
    1. Missing persons—Investigation—United States—Handbooks, manuals, etc.
    I. McGovern, Roger.   II. Title.
HV6762.U5G86   1996
362.8—dc20                                                              96-16787
                                                                            CIP

Printed in the United States of America

BOOKS ARE AVAILABLE AT QUANTITY DISCOUNTS WHEN USED TO PROMOTE PRODUCTS OR SERVICES. FOR INFORMATION PLEASE WRITE TO PREMIUM MARKETING DIVISION, PENGUIN BOOKS USA INC., 375 HUDSON STREET, NEW YORK, NY 10014.

*I dedicate this book*
*to the more than one million Americans*
*who disappear every year, many*
*of whom are never heard from again.*
*God bless them.*

# Acknowledgments

Any author who has been asked to update such a book as this, filled as it is with notoriously short-lived specifics—telephone numbers, addresses, names, and data—finds a certain amount of joy in so doing. It means that the original version performed as intended—helped a lot of people have the thrill and satisfaction of successful searches. The update would otherwise not have been in order.

The book itself exists as the result of my friendships with a number of knowledgeable and generously willing individuals who chose to contribute to what they felt was a worthwhile project.

These include R. J. and Sandy Smith of Dallas, there when I needed financial and moral support; my top investigator Judy Hanson; capable associates Diane Evans and Nancy Mindell; my capable attorney Robert S. Rose; superb investigator and researcher Scott Ross; attorney and friend F. Lee Bailey, who inspired me to go about this task; Robert and Roberta Duffy, always available for moral backup; associate Tom Lannin, Ph.D., University of California, Riverside, counsel in preparing the final manuscript; Janice Reed, Elizabeth

Williams, Aline Messer, Jennifer R. Moe, and Valerie Province for their splendid typing and editing skills; I thank also Marie Nelson, credit expert extraordinaire, whose wisdom can be seen in the body of the text; friend and fellow author Jerry Potter; P. E. Beasley and Don Kitchen, the two best "cops" I've ever known; the GDLG of the University of Nebraska; my most loved ones Greg, Lorie, Ted Jr., Mike, and their families; my sister Jo Ann and her family, and last—who should have been mentioned first—my beloved mother, Betty Gilliam.

# Contents

**Preface**   xiii

**1. Organizing Your Search**   1

ABOUT DANNY NOLAN  ■  MOTIVATING YOURSELF  ■  EX-
PENSES  ■  BEGINNING NOTES  ■  ABOUT PRIVATE INVESTI-
GATORS  ■  TWO YOU-CAN'T-DO-WITHOUTS

**2. Some Basic Search Tools**   14

U.S. POSTAL SERVICE  ■  THE HAINES CRISS-CROSS DIRECTORY
■  TELEPHONE DIRECTORIES  ■  PUBLIC LIBRARIES

**3. City Records**   20

PUBLIC LIBRARIES  ■  POLICE DEPARTMENTS  ■  CITY-COUNTY
PERMITS AND LICENSES

### 4. County Records   23

THE SHERIFF'S OFFICE ■ BIRTH, DEATH, AND MARRIAGE ■ REAL AND UNSECURED PROPERTY ■ FICTITIOUS BUSINESS NAMES ■ DISTRICT ATTORNEY'S FAMILY SUPPORT UNIT ■ GRANTOR/GRANTEE ■ CIVIL AND CRIMINAL COURTS ■ VOTER REGISTRATION ■ PUBLIC WELFARE ■ WORKMEN'S COMPENSATION ■ LICENSES AND PERMITS

### 5. State Records   42

DRIVER/VEHICLE RECORDS ■ CORPORATE RECORDS ■ STATE POLICE/HIGHWAY PATROL ■ STATE TAX BOARDS ■ PERMITS AND LICENSES ■ STATE SALES TAX BOARD

### 6. Federal Records   49

FREEDOM OF INFORMATION AND PRIVACY ACTS ■ SOCIAL SECURITY ■ THE U.S. POSTAL SERVICE ■ U.S. DISTRICT COURT ■ BANKRUPTCY COURT ■ MILITARY LOCATORS ■ MILITARY RECORDS ■ U.S. MARSHAL ■ INTERNAL REVENUE SERVICE ■ INTERSTATE COMMERCE COMMISSION ■ ALCOHOL, TOBACCO, AND FIREARMS ■ THE CENSUS TAKERS ■ GOVERNMENT PRINTING OFFICE ■ SELECTIVE SERVICE ■ FEDERAL OFFICE OF CHILD SUPPORT ENFORCEMENT ■ FEDERAL AVIATION AUTHORITY ■ PRISON RECORDS

### 7. Miscellaneous   62

EDUCATIONAL CHANNELS ■ NEWSPAPER PERSONALS ■ NEWSPAPER "MORGUES" ■ CATALOG MAILING LISTS ■ UNIONS AND ASSOCIATIONS ■ INSURANCE RECORDS ■ MAGAZINE SUBSCRIPTION LISTS ■ FYI—TELEPHONING FREE ■ THE SALVATION ARMY ■ RELIGIOUS AFFILIATIONS ■ *WHO'S WHO IN AMERICA* ■ HOME UTILITIES ■ THE TELEPHONE COMPANY ■ CHAMBERS OF COMMERCE ■ CLIPPING SERVICES ■ BOOK SOURCES

8. **Computers, Credit, and Consumers** 74

INTERNET ■ CONSUMER CREDIT REPORTING AGENCIES ■ CREDIT CARDS ■ BUSINESS CREDIT REPORTING AGENCIES

9. **Missing Persons** 81

THE VANISHING ACT ■ THE NAME CHANGERS ■ THE MISSING-PERSONS REPORT ■ THE NCIC AND MISSING PERSONS ■ MISSING CHILDREN ■ KIDNAPPING AND PARENTAL KIDNAPPING DEFINED ■ CONFLICTING KID-NAPPING ESTIMATES ■ TIPS FOR CUSTODIAL PARENTS ■ THE NATIONAL CENTER FOR MISSING AND EXPLOITED CHIL-DREN ■ MISSING CHILDREN HOT LINES ■ IF YOUR CHILD IS MISSING

10. **Safety Tips** 91

SAFETY TIPS FOR CHILDREN ■ FOR MORE INFORMATION ■ SAFETY TIPS FOR ADULTS

11. **The Adoptee/Birth Parent Search** 105

THE NEED TO KNOW ■ BEGINNING NOTES ■ SUPPORT— MORAL AND PRACTICAL ■ TRACING BIRTH PHYSICIANS ■ PETITIONING THE COURT

12. **Genealogical Library—The Church of Jesus Christ of Latter-day Saints (Mormon)** 122

THE FAMILY HISTORY LIBRARY

13. **Where to Write for Vital Records** 128

In Conclusion 130

**Appendix 1: Social Security Index of Valid Numbers** 132

**Appendix 2: Missing and Abused Children Organizations** 134

**Appendix 3: State Medical Boards** 137

**Appendix 4: Current Adoption Literature** 141

**Appendix 5: The Family History Library and Family History Centers** 143

**Appendix 6: Accredited Genealogists** 185

**Appendix 7: Addresses for Driver's License Transcripts** 198

**Appendix 8: Addresses for Birth, Death, Marriage, and Divorce Certificates** 203

**Appendix 9: State Adoption Records** 221

**Index** 241

# Preface

For almost all of my forty-five working years, my job has involved looking for people. People and evidence. But mostly for people important to an FBI case, or to one of my own. And at some point I became aware that almost everybody seems to be looking for somebody. So it occurred to me that it might be helpful to pass along a few of the search methods and information sources I've learned about, especially to those who can't hire somebody to do the looking for them and who can't go looking for themselves.

So I wrote this book in order to provide information about ways to search without leaving home base. It shows you how to reach more than 1,500 different, responsible, and usually cooperative sources of information on individuals through the use of the postal system and the telephone.

Chapters 1 and 2 offer guides on orientation, motivation, a short treatise on private investigators, and a look at some basic hunting tools.

The book is arranged with convenient records segments, so that you can pursue one specific search phase at a time: city, county, state, and federal. Because *that*—as they say—is where it's *at*; in the countless files that keep millions of bureaucrats employed full-time.

We have listed the files by name, spelled out the kind of information you can expect to get from each of them, and suggested ways to retrieve it with the least amount of contention.

Chapter 7, "Miscellaneous," gives your common sense a chance to come up with some of your own solutions. In Chapter 8, "Computers, Credit, and Consumers," we name the best of the personal information-retrieval services for a fee, and show how credit cards can sometimes give out information about those who own them.

I also have felt compelled to say something about "private eyes," the professional private investigators. Maybe it has crossed your mind, if you can afford it, to hire a P.I.—which brings me to one of the reasons I have prepared this book.

Investigation of the sort you want to pursue in locating a "lost" individual doesn't demand that you have a P.I.'s experience, but common sense helps a lot. No special skills are required; one logical step follows another. What I hope to do is organize that progression and help you develop the discipline that will keep your search efficient and make it productive.

I must emphasize that you should be flexible enough to accept help from unexpected sources and in unusual instances. The Salvation Army once gave backup to a New York City patrolman I know, who can still get round-eyed remembering something he witnessed at Grand Central Terminal.

A young woman and her six-year-old daughter were waiting on a mobbed platform to board a train to Yonkers after watching Macy's Thanksgiving Day parade. The train pulled forward for some reason, leaving the tracks beside the platform bare. The train was backing up again when the child suddenly fell onto the tracks and lay there stunned. The cop himself admitted that he was completely unable to move, frozen to the spot. But an instant before the child would have been crushed beneath the wheels of the train, a shabby old man leaped down from nowhere, seized her, and saved her from death. The man vanished into the throng after restoring the child to the platform and into the arms of her hysterical mother.

Each Friday until Christmas week, the mother returned,

traveling the thirty miles from Yonkers to search the crowds with my policeman friend, looking for her scruffy hero. The cop finally asked the Salvation Army for help. From the description of the man, by Christmas week the Salvation Army had a street name and an address for the shabby one. He was "Felix," a resident at the Salvation Army shelter in Greenwich Village. "Felix" was located and compelled not only to listen to the mother's tearful speech of thanks, but also to eat a magnificent breakfast at a fine restaurant under the policeman's watchful eye. (You'll learn more about the Salvation Army's helpful record-keeping efficiency later.)

None of the foregoing is typical of the search methods we'll pursue here. The point is that almost no one—not even the homeless—goes unrecorded by one agency or another. And that is the foundation on which this book is based: records. Records that can mean a successful search whenever there is a genuine desire to locate someone for whom you have either great affection or great need. You will find the tools for doing that—at minimal cost, for the most part—in the pages ahead.

Something else I think you'll find is a growing excitement as your search gets under way and clues and leads fall into place. This has been my experience. This is what has kept me fascinated for more than forty-five years as an investigator.

So be prepared for a little "rush" that comes with getting information you need out of a balky agency, discovering a clue that has been there all the time without your seeing it, the long shot that works out.

Also be prepared for the "down" days, when nothing goes quite right and you experience an epidemic of right names, wrong people; telephone numbers said to be valid turn out to have been disconnected for months; the person you've selected as your cheerleader and main support suddenly says, "C'mon, let's forget it." Don't you do it!

Nobody ever said being an investigator was easy. I've known searchers who got caught up in their need to succeed to the exclusion of almost everything else. You may not want to go that far. But if you begin to feel the pinch of frustration and sense that your main support is failing, act fast.

Sit down and rethink your entire plan and, if necessary,

select someone else as your lead supporter. Fire off a new batch of correspondence to sources you have not written before, but should have. Consult your supporters. Ask them for some fresh suggestions. Stop thinking negatively.

In any case, you haven't really started just yet. So let's explore a search in which I was involved, mostly as a consultant. It will give you some idea of the courses to pursue, the wide range of information sources available, and the kind of tenacity it sometimes takes to bring a search to a successful conclusion.

As you will see from the following, Danny Nolan's reason for initiating a search may appear to be something less than urgent compared to the one that caused *you* to pick up this book; the worst result of failure would have been Danny's profound disappointment. But regardless of the degree of urgency you're feeling, read Danny's story as the *anatomy of a search*.

# HOW TO LOCATE ANYONE ANYWHERE

# ORGANIZING YOUR SEARCH

## ABOUT DANNY NOLAN

Danny Nolan called me in Los Angeles as soon as his midterms were over and he knew he would graduate from the University of Nebraska in June.

"I want to invite my dad to come see me get my diploma."

Not so much to ask for, I thought, unless you knew, as I did, that Danny hadn't seen or really heard from his father, Tom, in more than a dozen years. Tom had left eighth-grader Danny and his mother, Marie, to shift for themselves in an unpaid-for house on Sioux Street in Lincoln, near my old neighborhood. Which is where I had come to know Tom Nolan, and why Danny was now calling me.

Tom was a financially ambitious fellow with an irresistibly attractive personality. Before disappearing, he had made some bad investments and borrowed a considerable amount of money. So for years he had been a subject of interest around Lincoln for others besides his son. A few unsuccessful attempts had been made by others to locate Tom. But love for a seemingly unworthy father—and *patience*—is what paid off for Danny when he decided to take up the search.

There are two sides to every story. One year after he disappeared, Tom Nolan began to send increasingly substantial monthly postal money orders to the house on Sioux Street. Only once in all those years did a message accompany the money. This agonizingly remorseful note arrived long after Danny suffered a high school football injury that put him in a wheelchair for life. But now the house was paid off, Danny had finished his education, he and his mother were comfortable. Tom also had apparently satisfied his creditors in Lincoln. What kept him away from his family? Whoever can answer such a question has the answer to a great deal of puzzling human behavior.

Marie didn't like the idea of Danny's search for his father, but she raised no objections. Now came the problem of locating Tom Nolan. It was at this point that Danny called me, and we spent a half hour discussing strategies.

The money orders were the only good clue we had. They all originated in Concord, New Hampshire. Because the money order sender is not required to supply an address, this was a "blind" lead. But Danny had his father's full name and birth date, which are required to retrieve information from any driver's licensing agency in the United States, so he wrote to the New Hampshire Division of Motor Vehicles. No luck, Danny reported to me. Tom Nolan, who had never been known to be without a car, was either not in New Hampshire or was driving under an assumed name, which wasn't likely. So the money order sender had to be just a helpful friend. Danny continued to follow my instructions.

A phone call to Tom's only surviving relative, a brother in Minnesota, produced the information that although Tom had attended their mother's funeral five years before, there had been no further family contact. But the brother had been able to add that Tom spoke at the funeral of his half ownership in a fishing boat in New Orleans.

The reference librarian at Lincoln's main library dug up the necessary addresses for the Louisiana Department of Safety (driver's license) and the Wildlife and Fisheries Department at Baton Rouge (fishing vessel licenses). Responses to letters to each informed Danny that (1) Tom Nolan's driver's license had expired, and (2) he had sold his share in the fishing vessel to his partner, whose name and address were supplied.

Response to a letter to the partner revealed that eighteen months earlier Tom had invested in an orange grove near Sebring, Florida. The partner had no address to offer. He wished Danny and his mother luck.

The cooperative reference librarian at Lincoln's main library found three Thomas Nolans in the Sebring area directory, none with the correct middle initial. The information operator in Sebring was of no help, so letters of inquiry, with stamped, self-addressed postcards, went to the Property Tax Office in Highlands County, where Sebring is located, and to the Department of Highway Safety in Tallahassee. Addresses were courtesy of the reference librarian.

The tax office postcard returned with a groaner: Tom had sold his orange grove to a citrus conglomerate, for some reason listing his old, useless New Orleans address. But a day later, a computer printout from Tallahassee brought pay dirt: Tom's nice, clean driving record—and his Florida address!

"Let's not risk a negative reaction to a phone call, let's *send* him the invitation," Danny suggested.

It took a week for the RETURN TO SENDER—NO FORWARDING ADDRESS stamp to bring the invitation back to the Nolans' mailbox.

"That's it. Forget it, Mom. This guy is a spook, and I don't think I want him haunting my graduation."

But Danny hadn't quite given up. One day, at the supermarket to which he sometimes accompanied his mother, he sat in his wheelchair at a magazine rack while Marie was shopping. He found himself flipping the pages of a publication called *Entrepreneur*, directed to people who want to be in business for themselves.

A tiny bell tinkled in the back of Danny's mind as he remembered something I had mentioned to him in running down the list of possible information sources. Maybe Danny could solve his own dilemma with an end around that just might net his enterprise-hopping father.

It took time to explain to Marie what he planned to do, and it seemed much too farfetched to her to work. But at my suggestion she wrote a touching note to Tom on an *unaddressed, stamped, forwarding postcard*, which she wouldn't let Danny read. And off it went, with a letter to *Entrepreneur*'s subscription manager. Now if Tom just happened to be on that subscription list, and if the subscription manager was a sympathetic type who appreciated their concern for Tom's privacy, maybe . . .

Admitting the remoteness of his chances for contact, with a resigned sigh Danny put away his search records, and with them his high hopes.

It didn't help when graduation day almost didn't dawn, the skies were so black with rain clouds. A radio newscast confirmed the expected. Commencement exercises, which had been planned for outdoors that year, would instead be held inside the Bob Devaney Sports Center.

There was standing room only as the wet-smelling audience settled down for the speeches and presentation of diplomas. From the platform, the graduates could hardly see the faces of people packed under the overhang at the rear of the hall. But then Danny wasn't looking—or was pretending he wasn't.

The ceremony was under way when something aroused the standees at the back of the room. The polite pushing and shoving attracted the attention of the graduates, including Danny. He saw a tall man shouldering his way through the crowd. It had been a long, long time, but Danny recognized the lean, handsome face he had given up hope of ever seeing again.

Tom Nolan. Adult runaway, check bouncer, reluctant family man, incurable entrepreneur, and in the end, loving father—home to make what amends he could.

As Danny Nolan's story exhibited here, and as we'll read more about in the pages ahead, knowing your subject's behavioral characteristics (along with his or her political and religious leanings) can be important to your search.

So can the degree of urgency you feel.

It isn't possible to imagine beforehand the degree of urgency, which is so important a search-motivation factor, for any of my readers. There would be a thousand variations. You, for example, could be facing a matter of life and death, perhaps needing to learn the genetic background of an adoptee who has become seriously ill. On the other hand, a legal or financial crisis might hinge on the success of your search. Then again, failure to find your subject might result in nothing more traumatic than disappointment.

In any case, I'm determined to get you into a frame of mind that gives your search *importance*. I want to give you something that, once your search has begun, should see you all the way through to a successful finish.

And that's what the following is designed to do.

## MOTIVATING YOURSELF

Cranking yourself into mental gear—motivating yourself—could be the most important element in getting your search under way. And of course it should come first, before you attempt any exploratory work whatsoever. So I suggest that you, as a beginning, do the following things.

- Read this book from cover to cover, *all* of it, not just the sections that seem to apply to *your* search.
- Give your search a priority ranking: casual, compelling, urgent, critical.
- Assign a beginning and completion deadline accordingly.
- Resolve to remind yourself often that, although this is a how-to book, you aren't building a barbecue pit or knitting a shawl for Aunt Olive. You're doing this for *you* and your feeling of well-being.
- Spend some time anticipating the pleasure and/or satisfaction to be derived from completing your project successfully despite frustrations and setbacks; there *will* be some of these.
- Accept my word for it; this is a fascinating game, one that you should convince yourself you're going to win.
- Look forward to the challenge to your patience, ingenuity, and—this will get a good workout—common sense.
- Enlist the support of family and friends. Their interest will help sustain *your* interest. Their suggestions will be invaluable.

- Plan to keep accurate records and a *journal* to which you can point with pride when your project is completed.
- From this page forward, think *success*!

## EXPENSES

The cost of getting your "friendly" living-room search off the carpet will be modest, probably somewhat less than the more complex searches covered in Chapter 8, "Computers, Credit, and Consumers," and Chapter 11, "The Adoptee/Birth Parent Search."

Here are some of the expenses you'll incur in your "friendly" search.

### MAILINGS

Ream of white writing/typing paper
Box of inexpensive white envelopes
Fifty (50) first-class postage stamps
Fifty meter-stamped postcards
Photocopying
Local travel (fares, gas)
Highlight pen
Lined legal pads (2)
  TOTAL: under $50.00

### TELEPHONES AND THE THREE-HOUR SPREAD

Consider the time of day in keeping down long-distance costs. Calls made on weekends between 5:00 P.M. Friday and 5:00 P.M. Sunday, and after 5:00 P.M. weekdays in your zone, offer a discount. Be aware that 5:00 P.M. Eastern is 4:00 P.M. Central, 3:00 P.M. Mountain, and 2:00 P.M. Pacific, with the business day still under way in those zones. A 5:00 P.M. California call, then, should find a New Yorker at home with dinner over. A call from New York at 11:00 P.M. reaches a West Coast callee at just about that same stage. So act like a calculating clock watcher and you won't find telephoning outrageously expensive. Remember—a call to the right party at the right time could abbreviate your search.

### DOCUMENT AND OTHER FEES

Some document fees are reasonably consistent across the country, such as driver's license records. Other fees, even for essentially identical documents, vary so much from state to state that I can't even offer estimates. This is more of a problem for those pursuing birth parent and relinquished-child searches. They can expect to pay not only for basic correspondence and telephone costs but for a clutch of documents as well. They usually need copies of birth, marriage, and death certificates and other records that require photocopying and notarizing. Their searches might involve mass mailings to a host of people with the same surname as the subject's. Then there could be membership dues to support groups, advertisements in—and subscriptions to—search publications.

The upbeat side of all this paperwork is that you can keep within your budget by doing some commonsense planning. For example, never send for a document unless you are certain it is absolutely vital. Write the agency beforehand for search and copying costs so there are no surprises. If the need for a mass mailing occurs, use less expensive double postcards rather than a letter with a stamped, forwarding postcard enclosed. Don't be put off by needing to run an advertisement or join a support group. Search publications charge modestly for both ads and subscriptions. Support and information group membership dues average only $20 to $30 per year.

Above all, keep good cost records; they will help guide you in the least costly way to go. You'll get into the habit of economizing.

## BEGINNING NOTES

### STEP 1

Use the Missing Person statistics list on pages 8–9 as a guide for creating a Profile of your subject. Then knock on doors. Telephone. Talk to former fellow workers, friends, enemies, lovers, relatives. Ask traditional who-what-when-where-why-how questions and take notes—mental or otherwise—of the answers.

Your objective here is to create as complete a word picture of your subject as possible. So add anything I've missed that you think should be part of his or her physical description and background.

Your subject's Profile is a document vital to an organized search. It will become part of your Inquiry Kit, which I will tell you about a little later. Keep adding details to your Profile as your search progresses. You'll be surprised (and your subject may be, too) at the amount of information you'll be able to collect.

Make at least two dozen photocopies of the Profile, one for each kit you'll assemble (you may need more). File the original Profile for further copying as needed. Interfaced (matched) with data for others with an identical name, it will help distinguish your subject from them.

### PROFILE OF SUBJECT

- **Full name**   First, middle, last, maiden, confirmation, and nicknames
- **Gender**   M/F
- **Vital statistics**   Day, month, year, place, of birth; Social Security number
- **Physical**   Height, weight, eye/hair color, glasses, facial hair, accent, lisp, tattoos, deformity, limp, scars, moles, etc.
- **Descent**   Caucasian, Black, Asian, Latin, Middle Eastern, Polynesian, etc.
- **Last address**   And the two before that, if available
- **Education**   Grammar, high, college, trade, years attended, graduated
- **Occupation**   Business, trade, profession, last employed as, retired, etc.
- **Organizations**   Labor, social, service, trade, professional, etc.
- **Religion**   Denomination, tither, frequent service attender
- **Military**   Service, period served, where, when, rank at discharge, serial number
- **Licenses**   Driver (state, number), pilot, barber, contractor, lawyer, doctor, etc.

- **Hobbies**   Fishing, knitting, hang gliding, golf, etc.
- **Search reasons**   Legal/health crisis, friendly/loving desire for contact, financial/credit
- **Relationship**   Birth parent/adoptee, immediate family, other relative, lover, friend, priest, doctor, lawyer, etc.
- **Subscriptions**   *People*, *Sports Illustrated*, *Aviation*, etc.
- **Possible location**   Where you've been told your subject is, where you think he or she is, where your common sense tells you he or she might be

### STEP 2

Much of your search will be conducted through the U.S. Postal Service. Lay in some standard $8^1/_2$" × 11" white typing paper; #10 envelopes; a highlight pen (for highlighting certain Profile items—military service, for example, when contacting the Veterans Administration).

Lay in a batch of meter-stamped postcards. And depending on how fancy a working journal you think fits the project, you might find a simple, lined legal pad quite adequate. Keep in mind that it is easy to lose track of details as you go along because your search is not something you'll be working on every day. There will be stretches of time when you're waiting for responses, and your memory of what you were doing last will start to fade. Your journal with notes of everything you've done up until that last working day will always be there for you to go back and pick up from when the waiting is over. Remember, *date entries* as you go.

### STEP 3

Get a membership card from your *main* (if there is more than one) *library*. Acquaint yourself with the *reference librarians* there, and explain your project to them. Librarians are almost invariably knowledgeable and willing to guide you through their world of facts, figures, dates, and place-names. They have the addresses of most of the important information sources you'll be contacting, plus the addresses of every other library in the United States. Knowing these details now will be important later.

### STEP 4

Use your legal pad/ledger or—if yours is a more complicated search—separate 3" × 5" filing cards for keeping records of contact names, phone numbers, addresses, dates, and other data. Make progress notes for each of your activities so you will be able to pick up easily where you left off even after long periods of inactivity. *Discipline* yourself. Be neat and accurate, as though your files might someday become a court record, which could well happen if yours is a search for legal reasons.

### STEP 5

Again, I suggest that you read *all* of this book before beginning work. Be aware that Chapter 8, "Computers, Credit, and Consumers," and Chapter 11, "The Adoptee/Birth Parent Search," can be helpful in all types of searches.

From this moment on, keep in mind that most successful searches result from ten vital attitude factors:

- enthusiasm,
- patience,
- a sense of dedication,
- discipline in your work,
- willingness to accept suggestions,
- perseverance,
- retaining a mental image of success,
- reliance on your common sense,
- the commitment to make things happen when it seems nothing is about to,
- acceptance of results if you succeed.

This last is the key to finding peace and satisfaction if you are looking for a relinquished child or a birth parent and are successful. Without first resolving to accept these consequences, whatever they might be, it may be best not to pursue your quest.

With all this out of the way, we are almost ready to go.

But first I want you to meet someone who could supply you with an alternative to doing your own searching.

## ABOUT PRIVATE INVESTIGATORS

You probably don't want to—or can't afford to—hire a private investigator to do your search for you; that's why you bought this book. But let's take a quick look at him or her and his characteristics so you can relate to him somewhat.

To begin with, P.I.s are easy to find. There are thousands, licensed and unlicensed, at work in the United States. Very few of them are real professionals. Of course, some are trained operatives with law-enforcement backgrounds who are completely dependable.

Private investigators are listed in most of the Yellow Pages between New York and Honolulu. The problem is choosing an effective one—unless you personally know of a capable P.I. you can expect effectively to carry out your assignment without its costing you an arm and a leg. An average fee for a P.I.'s services, whether he's good or not so good, can run as high as several hundred dollars a day, with the inevitable "plus expenses," an item that can become financially lethal over a long search.

In this book, I've discussed some of the techniques your eye-for-hire would normally use. But for the most part, although the P.I. depends a lot on personal contact with his information sources, you will be using the mails, the telephone, and public records.

Another major difference between you and some paid sleuths is their willingness to wink at the law, using illegal acts to help shortcut the gathering of evidence; you've learned a few of these from watching television. And of course, it's not only illegal but unwise to represent yourself as a law-enforcement officer or government official at any time or for any reason. This kind of activity can get you into very hot water.

So my firm's policy with clients is: "We'll investigate anything and anybody, but we won't go to jail for you."

However, there are legal ruses we're not above using

when the end justifies the means and we feel that gaining the information you need is reason enough. You might identify yourself on the phone as a telemarketer/surveyor wanting to know your subject's present occupation, or you might say you're a friend of old hometown friends with a message from "the gang" back there. This kind of subterfuge is not illegal; just don't try to represent yourself as a government official.

Beware, too, of coming off as stupid by asking the wrong questions, or by being unsure of yourself. Anticipate questions about your identity and have the right answers ready.

If you're into an investigation that calls for subterfuge and your subject is obviously trying to avoid you and detection, leave him or her alone. Look up the subject's relatives and pay *them* a visit. Or locate a girlfriend or boyfriend, maybe former workmates. Forgo any Sam Spade stuff. Minor deceptions here and there will gain you what you need to know.

## TWO YOU-CAN'T-DO-WITHOUTS

Before you begin any kind of comprehensive search, you must possess *two* vital statistics regarding your subject.

1. *Your subject's first, middle, and last names* (correctly spelled), the name he or she is licensed by, votes by, and gets credit with, whether it is a birth or adoptive name, and nicknames, if any.

If a birth or adoptive name has been changed *officially*, your identification problem is heightened and may call for a search of court records or for interviews with associates who know the subject by his or her new identity. If the subject has *assumed* a different name, an aka (also known as), that complication will also slow the progress of your search. These are all solvable problems if you read this book thoroughly.

2. *Your subject's date of birth* (DOB). Because of a population nearing the 250 million mark, there are thousands of name duplications in the United States. Of all the search data you can supply to information sources—in addition to a full

name—the DOB is the clearest identifying statistic separating your subject from others of the same name. (The place of birth [POB] is another helpful identifying item you should try to acquire.) If you don't already have it, you'll learn from this book how to go about getting the DOB.*

---

*Less critical, but helpful in your search, is knowing your subject's Social Security number. The Social Security number is used in many ways as an American serial number.

# SOME BASIC SEARCH TOOLS

## U.S. POSTAL SERVICE

Most government agencies—and private ones, too—are like some people: they hate to divulge secrets, even if the secrets are not their own. "It's for me to know and you to find out" is their attitude. It's a good thing, too, for by retaining the right of privacy for you and for me, agencies protect us from irrelevant prying.

We are not going to challenge that right, nor are we going to pry in any sense. But our problem here is to make the agency person quickly aware of (1) why we want to contact a subject, and (2) how we can achieve this goal without invasion of the subject's privacy.

You can use a mailing device that will allow you to make contact without arousing suspicion or resistance in those guarding citizens' privacy.

Here is how it works; it's very simple.

■ Determine which public or private agency is likely to have your subject's current address, or a source for the address:

Veterans Administration, Social Security, a state vital statistics office, an insurance company, whatever.

- Write a short, clear *letter of inquiry* addressed to either the director or the supervisor of the agency (see directly below).
- Enclose a photocopy of your subject's Profile, highlighting statistics that might apply to that agency: military service, for example, when contacting the Veterans Administration.
- Enclose a *plain, post office meter-stamped postcard* (see page 16) on the back of which is a brief, clear message to your subject

---

**LETTER OF INQUIRY TO THE VETERANS ADMINISTRATION**
Director, Regional Office
Veterans Administration
(your regional VA office address)

Regarding: Charles Thomas Ingham
          Date of Birth: 9/14/59
          In Louisville, Kentucky

(Date)

Dear Director:

    I have an urgent reason for contacting the above-named former serviceman: U.S. Army, 1990–1993, Persian Gulf Theater, approximately 10/1/91 to 8/1/93. Discharge rank: sergeant. I have enclosed a vital statistics profile. If his present address is in your file, I request that you please forward to him the enclosed, stamped, unaddressed postcard. If his address is not available, please have the postcard mailed back to me.*

    Thank you for your help.

Peter J. Anderson (your signature)

Peter J. Anderson (typed or printed)
107 Maple Terrace
Montgomery, AL 36104

---

*Note*: Always request that the forwarding postcard be mailed back to you if your subject's address is not available. This will help you keep accurate files and eliminates the agency or individual as an information source.

asking him or her to contact you; your name, signature, return address; and—if desirable—your telephone number. Leave the front (stamped) side of the postcard blank, so that your subject's name, if available from the agency's rolls, can be applied for forwarding. You now have—with your letter of inquiry; Profile; and stamped, unaddressed, forwarding postcard—what we will call from now on your Inquiry Kit. (An enclosed photograph of the subject would help.)

*Note*: If your longhand is as illegible as mine, I suggest that you print all your correspondence rather than use a typewriter. Pen and ink tends to attract the attention of clerks and officials exposed daily to bales of typewritten material. Furthermore, pen and ink will enhance the personal nature of your inquiry—and of the postcard to your subject.

Also, ask for the names and phone numbers of a couple of your subject's former neighbors. They just might be nosy enough to have learned where your subject was going when he or she left town.

You may have a same-name problem; Browns, Smiths, and

FORWARDING POSTCARD

RETURN ADDRESS

U.S. POSTAGE
STAMP

Leave front blank for your subject's address.

Dear Chuck: I asked the Veterans Administration to forward this postcard, since I have no idea where you are and we need to talk. I've got good news for you. Please write, or call collect.

Peter J. Anderson
107 Maple Terrace
Montgomery, AL 36104

If the postcard is returned to you, take a deep breath and try again somewhere else.

---

Williamses take up dozens of pages in metropolitan directories. But if you don't have this problem, and want to do a thorough directory search, don't hesitate to call all the numbers listed under your subject's surname. You could well reach a relative who knows exactly where your subject is. Just be prepared to fib a little, if necessary, about your reasons for wanting to reach the subject; families don't give up the location of a relative to strangers all that easily. You might want to say you're the subject's former insurance agent and that you have some cash residue from an old policy you want to return to the subject. That should get you at least an address to work with, and maybe a phone number.

## THE HAINES CRISS-CROSS DIRECTORY

The Criss-Cross is a cross-reference directory. It lists every street in any given community alphabetically, from Abbey Lane to Zoeotrope Circle, and gives the names, addresses in sequence, and telephone numbers (except for unlisted numbers) of the people who live on those streets.

That's the first half of the Criss-Cross. The second half lists all the telephone numbers, in sequence—with the prefix as a base—and the street names, addresses, and residents. So if you have nothing to go on but a telephone number, the

Criss-Cross will let you run it straight down to the address and name.

## TELEPHONE DIRECTORIES

Once in a while I get calls from people in Greater Los Angeles who are quite serious about hiring me to find someone without their first checking telephone directories for southern California. They ignore a rule of common sense. More than 6 percent of the nation's population—some 14 million people— live within Greater L.A.'s area codes 213, 310, and 818. Because I begin all searches by starting from ground zero, occasionally I have been able to respond to search callers on the same day with a confirmed address I've found in a phone book here in my office. Directories can be helpful tools and often are overlooked.

But if you've already checked local directories for your subject with no luck, try your public library, which has a collection of current phone books for various regions of the United States. (Telephone companies used to provide libraries of directories for every area of the United States at their main offices, but they don't any longer.) At your main library you'll probably find a directory for the community in which your subject last lived.

OK—you found the right directory, but not your subject. Then try this: almost all public libraries also maintain a file of *local* phone books for past years—up to twenty years or more in some cases. Write or call the reference librarian in your subject's suspected area and ask for a check of local directories over the past three or four years. If his or her name pops up in one, ask the librarian to run a check in her Haines Criss-Cross Directory for who's living at your subject's old address *now*.

## PUBLIC LIBRARIES

You probably already have a library card. If you don't, get one, if only to be on the library's records. Most of the material you'll be looking up will be in the Reference Room, and whatever is there stays there. It can't be checked out.

Be aware that your main library—if your town has more than one—is probably better equipped with reference material than the branch libraries.

Go out of your way to be pleasant and appreciative to the reference librarians. They are absolutely invaluable in helping you with search information.

# CITY RECORDS

## PUBLIC LIBRARIES

Long experience with the most marvelous free public library system in the world has led me to a dependency I hope *you* acquire. Many reference librarians seem to know everything about everything, or have answers at their fingertips in volumes of exotica or minutiae that make them fountains of information. These people are—without exception, in my experience—not only incredibly patient but blessed with understanding and an aversion to your leaving their rooms with your information problem unsolved. Take advantage of this extraordinary service. It's free.

I bring up the subject of libraries often because I think so much of them and the people who run them. At this time I also want to mention a source of information they have that isn't related to the books on their shelves.

It's the collection of cardholder names they have in their computers. I doubt very much that these dedicated people would release this kind of information to the public. But they would regard helping you make contact with someone on their cardholder list as part of their job, perhaps by sending a card regarding your search. There are thousands of public libraries throughout the country, most of them with computerized record systems, giving you still another way to go in your search. We'll discuss Internet later.

Get in touch with the library system in the area where your subject may be living, and put your Inquiry Kit to work.

I haven't mentioned before, but I do now, that there is a *law library* in just about every American courthouse, or in a building nearby. Most law libraries maintain well-cataloged and updated state and federal literature on adoption, child relinquishing, child abuse, kidnapping, parental abduction, foster parenting, and so on, free for you to use.

*University and college libraries* offer—aside from massive collections of textbooks and classics—the results of special studies and research directed to family relations and child abuse. The library system of the University of California's campuses alone (eight campuses total, excluding the Medical/Dental Center in San Francisco) contains over 19 million volumes! Even the smallest campus, that of UC-Riverside, has several libraries holding over 1.25 million bound and unbound texts, including tens of thousands of rolls of microfilm.

All of the library services I've mentioned here are free. And again I remind you that the list of cardholders at any of these libraries may be a source of information to you—if you use your Inquiry Kit properly.

## POLICE DEPARTMENTS

Police departments and all law-enforcement agencies are tied into terminals for the for the National Crime Information Center (NCIC) computers in Washington, D.C. You can now ask your police department to list a missing child with NCIC just a few alarming hours after his or her disappearance. It soon may be possible to do the same with missing, mentally capable adults if planned modifications are made in allowable entries into the NCIC system. (Only retarded and aged adult missing had been eligible for NCIC entry up to 1988.)

The typical police department maintains extensive records. It keeps a file of local known active lawbreakers and nuisance perpetrators. And a twenty-four-hour running account of who is doing what to whom comes over the statewide teletype network that reaches all law-enforcement agencies as well as the NCIC. This flow of crime information can be interrupted at any time by any agency so that it can enter its own report of local criminal activity, with physical descriptions of suspects, modus operandi, and vehicles operating in that agency's jurisdiction.

But the task of massive criminal recordkeeping is pretty much up to the county and the state. The exceptions are monster population centers such as New York, Chicago, and Los Angeles. Because of the sheer volume of lawbreaking activity in these areas, big-city police departments must share in the jailing and the recordkeeping.

In most areas, however, as prosecution and disposition of wrongdoers progresses, the county and state take up the burden of keeping track of them. At another level, the federal court and penal systems, and law-enforcement agencies such as the FBI, pick up recordkeeping and jailing for their special categories of lawbreakers.

Just keep in mind that, thanks to the Freedom of Information Act (see page 49), you can open many of the records that once were closed to the public.

## CITY-COUNTY PERMITS AND LICENSES

Cities and counties regulate almost all business and civic activity within their boundaries by issuing permits. These permits include the name and address of the individual applying. Such records are normally kept at City Hall and the county courthouse. They would include permits for building projects, demonstrations and parades, as well as dog licenses, bicycle licenses, vendor licenses, and so on. City and county business licenses are located in the city and county clerks' office. Just about all businesses operating legally in cities and counties are required to have a license. The application for the license contains the name and address of the business, the name and address of the owner or his or her agent, the owner's home phone number, the type of business, the number of employees, the date the application was filed and the license's expiration date, and the filing fee paid.

# COUNTY RECORDS

Any American county courthouse is a gold mine of records concerning the people who live there and the legal activity that takes place between them and the elected authorities.

Many of these records can be obtained simply by knowing the subject's full name and date of birth. Of course, you should have a good idea of what *kind* of record you want before you can determine which office and what clerk to contact—by telephone, by mail, or in person.

What follows is a discussion of some of the county records available for search and the range of information each of them provides.

## THE SHERIFF'S OFFICE

The sheriff is elected, not appointed. In most counties, he is considered the number-one law-enforcement officer. (Some counties give this ranking to the district attorney.) The sheriff's deputies serve subpoenas, warrants, sheriff's sale notices, and handle sale details. They patrol unincorporated highways, investigate motor vehicle accidents on private property, investigate crime (with a detective division), operate the jails, provide guards for the courtrooms (bailiffs), and transport inmates from one facility to another.

And, as if all that didn't keep them busy enough, the sheriff's staff must keep extensive jail records.

Prisoner files will include mug shot, full name, akas, next of kin, recent addresses, fingerprints, charges, jail conduct, visitor list and dates, unusual correspondence (as a rule, mail is censored), statements made while in jail, trial dispositions, and much more.

These records are hard to get to, but you can access them with formally prepared requests through the district attorney and—with considerably more difficulty—through the public defender. Because the Sheriff's Department and the district attorney are the prosecutors, the public defender is necessarily an adversary to them. And because both sides are handling admissible trial material, they both tend to guard their records jealously. Of course, the defender's inventory of records at any point in time is minuscule compared to the sheriff's.

In any case, don't be too hopeful of retrieving much information from either of these sources through the mail.

Be prepared to appear in person (1) with a lawyer at your side and (2) with a good knowledge of procedure and very persuasive ways. If you can't appear in person, and a truly urgent need exists for special information, you can write to the court involved.


## BIRTH, DEATH, AND MARRIAGE

### BIRTH RECORDS

Copies of original birth certificates are routinely supplied on request to everyone except an adoptee, whose *original* certificate is sealed by the court on the day the adoption is final and can be opened only by court order (see page 25). The adoptee is allowed a copy of his *amended* certificate, from which all of the "identifying" information has been eliminated. These deletions include the names of the natural parents, their places of birth, their ages, the name of the certifier (usually the attending physician), and the name of the registrar for the county. In short, nothing that might help lead a searching adult adoptee to his true origins is included in the amended birth certificate.

Except for two items.

Each newborn is given an *official number* that is filed by *date of birth* in the county registry. Almost every county retains those two items *unchanged* on the amended birth certificate. So, if a searching adult adoptee knows his exact birth date and the number from his amended certificate, a matchup in the birth records could reveal his natural birth name. These things can help in identifying his birth parents.

Birth records are maintained at both the county and the state vital statistics offices. They are public information—except for adoptees' original birth certificates—available to you in person or by mail with submission of the proper request. Write for a fee figure and include a check or money order with the actual request.

### PROVIDE THE FOLLOWING FACTS WHEN REQUESTING A BIRTH CERTIFICATE

- Subject's full name
- Gender and race
- Subject's parents' names, including maiden name of mother
- Date of birth: month, day, year
- Place of birth: city, town, county, state, hospital
- Purpose for which certificate copy is needed
- Your relationship to person whose record is being requested (See Appendix 8 for addresses in each state to use when submitting requests for birth certificates by mail or in person.)

### Birth Records of Persons Born in Foreign Countries Who Are U.S. Citizens at Birth

Births of U.S. citizens in foreign countries should be reported on the Consular Report of Birth (Form FS-240) to the nearest American consular office as soon after the birth as possible. This report should be prepared and filed by one of the parents. However, the physician or midwife attending the birth or any other person having knowledge of the facts can prepare the report.

Documentary evidence is required to establish citizenship. Consular offices provide complete information on what evidence is needed. The Consular Report of Birth is a sworn statement of the facts of birth. When approved, it establishes in documentary form the child's acquisition of U.S. citizenship. It has the same value as proof of citizenship as the Certificate of Citizenship issued by the Immigration and Naturalization Service. Filing a Consular Report of Birth is not authorized for children five years of age or older.

A modest fee is charged for reporting the birth. The original document is filed with: Passport Services, Correspondence Branch, U.S. Department of State, Washington, DC 20524. The parents are given a certified copy of the Consular Report of Birth and a short form, Certification of Birth (Form DS-1350 or Form FS-545).

To obtain a copy of a report of the birth in a foreign country of a U.S. citizen, write to the Passport Services office. State the full name of the child at birth, date of birth, place of birth, and names of parents. Also include any information about the U.S. passport on which the child's name was first included. Sign the request and state your relationship to the person whose record is being requested and the reason for the request.

The fee for each copy is $6.00. Enclose a check or money order made payable to the U.S. Department of State.

The Department of State issues two types of copies from the Consular Report of Birth: (1) a full copy of Form FS-240 as it was filed, or (2) a short form (Form DS-1350 or Form FS-545), which shows only the name and sex of the child and the date and place of birth. The information in both forms is valid. The short form may be obtained in a name subsequently acquired by adoption or legitimation after proof is submitted to establish that such an action legally took place.

### Birth Records of Alien Children Adopted by U.S. Citizens

Birth certifications for alien children adopted by U.S. citizens and lawfully admitted to the United States may be obtained, if the birth information is on file, from: Immigration and Naturalization Service (INS), U.S. Department of Justice, Washington, DC 20536.

Certification may be issued for children under twenty-one years of age who were born in a foreign country. Requests must be submitted on INS Form G-641, which can be obtained from any INS office (addresses can be found in a telephone directory). For a Certification of Birth Data (INS Form G-350), a $5.00 search fee, paid by check or money order, should accompany INS Form G-641.

Certification can be issued in the new name of an adopted or legitimated child after proof of an adoption or legitimation is submitted to INS. Because it may be issued for a child who has not yet become naturalized, Form G-350 is not proof of U.S. citizenship.

## Certificate of Citizenship

United States citizens who were born abroad and later naturalized or who were born in a foreign country to a U.S. citizen (parent or parents) may apply for a certificate of citizenship pursuant to the provisions of Section 341 of the Immigration and Naturalization Act. Application can be made for this document in the United States at the nearest office of the Immigration and Naturalization Service. The INS will issue a certificate of citizenship for the person if proof of citizenship is submitted and the person is within the United States. The decision whether to apply for a certificate of citizenship is optional.

## Records of Births Occurring on Vessels on the High Seas or on Aircraft in International Flight

When a birth occurs in international territory, whether in an aircraft or on a vessel, the determination of where to file the record is decided by the direction in which the vessel or aircraft was headed at the time the birth occurred.

1. If the vessel or aircraft was outbound or had docked or landed at a foreign port, requests for copies of the record

should be made to: U.S. Department of State, Washington, DC 20520.

2. If the vessel or aircraft was inbound and the first port of entry was in the United States, write to the registration authority in the city where the vessel or aircraft docked or landed in the United States.

3. If the vessel was of U.S. registry, contact the U.S. Coast Guard facility at the port of entry.

### Records Maintained by Foreign Countries

Most, but not all, foreign countries record births. It is not feasible to list here all foreign vital records offices, the charges they make for copies of records, or the information they may require to locate a record. However, most foreign countries will provide certifications of births occurring within their boundaries.

U.S. citizens who need a copy of a foreign birth record may obtain assistance by writing to: Office of Overseas Citizens Services, U.S. Department of State, Washington, DC 20520.

Aliens residing in the United States who seek records of these events should contact their nearest consular office.

#### DEATH RECORDS

There is the possibility that your subject may no longer be alive (see pages 29–30 for data needed when requesting death records). If he or she is advanced in years, and you have no idea where he or she resided last, this is more than a possibility, and a good first move is to contact the Social Security office (see page 51). Under ordinary circumstances, Social Security is among the first of the agencies to be notified of a death, usually by the coroner or by the next of kin.

But some deaths *aren't* ordinary, and you should be aware that this notification can be made only if the deceased is found bearing comprehensive identification: for example, a Social Security card (which many older persons no longer carry, having memorized the number long ago), a driver's license, or credit cards. But in many cases of homicide, suicide, and accidental and unattended death, the possibility of the deceased's being separated from his or her ID after death is a strong one. If a body is discovered without positive ID, the following procedure takes place in most counties of the United States.

The coroner attempts to determine the victim's identification through postmortem examination, referring to dental charts, fingerprints, X rays, tattoos, scars, location of the body, and other details. If all fail, the decedent's death particulars go to what is called, in California, the County Public Administrator's Office (CPAO), which has a counterpart in most American counties. The CPAO turns over whatever is found with the victim to the county treasurer or some general fund. Burial then takes place at the county's expense.

If the deceased is identified but remains unclaimed, the name is entered as a public record in the state's vital statistics office. Any researcher can usually procure a copy of the death certificate and details of the death by mail, in person, or by contacting this office. Such an office may also be maintained by the county in which the decedent resided. Check it.

When the victim is known and claimed, the death certificate is normally prepared by the attending physician, funeral director, hospital authority, or coroner. Then it is filed in both the county in which the death occurred and at the state's vital statistics office.

The certificate is public information, for a fee, by mail or in person. It will probably contain the deceased's birth name, age, date and place of birth, last known address, physical description, injury description (if any), laboratory results of biopsy/autopsy and cause of death, names and addresses of informants, next of kin, funeral director, cemetery interred, or crematory. Also listed (if available): the deceased's occupation, military record, and other life details.

This is the same information you would be asked for if you were reporting a death to authorities.

**SUPPLY THE FOLLOWING DATA**
**WHEN REQUESTING DEATH RECORDS**

- Full name of decedent
- Gender and race

- Parents' names, including maiden name of mother
- Date of birth: month, day, year
- Place of death (if known): city, town, county, state
- Purpose for which certificate copy is needed
- Your relationship to the deceased
  (See Appendix 8 for addresses in each state to which you should submit requests for certificates. Write for the exact fee figure beforehand, and be sure to include it when you mail your request.)

### Death Records of U.S. Citizens Who Die in Foreign Countries

The death of a U.S. citizen in a foreign country is normally reported to the nearest U.S. consular office. The consul prepares the official Report of the Death of an American Citizen Abroad (Form OF-180), and a copy of the report is filed permanently in the U.S. Department of State (see exception below). To obtain a copy of a report, write: Passport Services, Correspondence Branch, U.S. Department of State, Washington, DC 20524.

*Exception:* Reports of deaths of members of the armed forces of the United States are made only to the branch of the service to which the person was attached at the time of death—army, navy, air force, marines, or coast guard. In these cases, requests for copies of records should be directed as follows. *For members of the army, navy, or air force:* Secretary of Defense, Washington, DC 20301. *For members of the coast guard:* Commandant, P.S., U.S. Coast Guard, Washington, DC 20226.

### MARRIAGE RECORDS

The standard *application* for a marriage license is a pretty demanding document (see page 31 when requesting marriage records). It usually requires full names, dates of birth, ages, number of marriages, date of last marriage, ending date of last marriage (and whether ended by death, divorce, or annulment), birthplace of former spouse(s), groom's present address, present and last occupation, kind of employment, highest grade completed, groom's father's name, mother's maiden name, and both of their birthplaces. That's just the application.

Some of these data have to be duplicated on the marriage *certificate*, which will also show the type of ceremony (civil or church), the bride's and groom's religion, the name of the person who performed the ceremony, and the names of the witnesses.

So marriage records offer many bits and pieces of information on people and places. The official records are filed in the county in which the license was issued, and in some states in the vital statistics office at the capital. These records are normally available on request—unless the couple asks that they be *sealed*.

### SUPPLY THE FOLLOWING FACTS WHEN REQUESTING MARRIAGE RECORDS

- Full names of bride and groom (include nicknames)
- Residential addresses at time of marriage
- Ages at time of marriage (or dates of birth)
- Month, day, and year of marriage
- Place of marriage
- Why you need the marriage information
- Your relationship to the married couple
  (See Appendix 8 for addresses of issuing offices—and don't forget to include the fee with your request.)

## REAL AND UNSECURED PROPERTY

Real property is normally administered at the tax assessor's office, which could be either a city or a county function and is a good source of information because these are public records.

In most states, real property means real estate—homes, lots, business buildings and equipment, farmland, land of any kind.

If your subject owns real property, he or she is certain to be found on the tax rolls. The tax bill will provide you with a mailing address and the location and value of the property. If the property has been sold to an absentee owner, the public tax records will provide you with the new owner's name and business address. You can probably get from this person the information you need about your subject. Such information also is likely to be available at the local Board of Realty office—if yours is a local search. Any friend with a real estate license can get you what you're looking for. All you need is your subject's county location and payment of the usually modest fee.

Unsecured property is administered by the same tax assessor. But the goods are in another category: for example, cars, boats, airplanes, mobile homes, office furniture. Get airplane ownership by contacting the FAA (see page 60 for the address); cars, boats, and mobile homes are usually registered with the Department of Motor Vehicles. The tax assessor's records tell where the property is and how much it's worth. Everything mentioned thus far is public information.

Simply write the tax assessor in the appropriate county.

## FICTITIOUS BUSINESS NAMES

If your subject's business is using a name other than his or her own—a fictitious name—he or she must have filed a statement to that effect. Without that document on file, banks are forbidden to open an account for an entrepreneur in any other than a birth name.

The fictitious name files offer the business owner's name, fictitious business name and address, and home address; associates' names and addresses; whether the business is incorporated and in what state; ownership arrangement (husband and wife, individual, co-partners, general partnership, or joint venture).

You may learn more about your subject from the fictitious name files than you want to know. They are open records at the county level, and the information is available by mail for a nominal fee.

## DISTRICT ATTORNEY'S FAMILY SUPPORT UNIT

Anyone filing for action against a parent who is defaulting on support payments can get help from the D.A.'s Family Support Unit. Its job is to go after dependent child and spousal support from absent parents. When the question of paternity comes up, the D.A.'s group will help resolve it. If there is no court order for child support, it will try to get one.

In some counties, a department other than the D.A.'s may handle family support. Your mail will be forwarded to that office.

There is a small problem involved. The files dealing with family support are generally confidential, and in most cases can be retrieved only by the people involved.

## GRANTOR/GRANTEE

Grantor/Grantee records contain transaction documents of every description, but may operate under a different name (in Allegheny County, Pennsylvania, the same administrator is called the prothonotary—a fancy Latin name for chief clerk). A prothonotary once rode in president-elect Harry Truman's car in a political parade. When told who his backseat companion was, Mr. Truman asked, "What the hell is that?"

Anyway, every Grantor/Grantee's office is overloaded with agreements, trust deeds, powers of attorney, judgments, and so on. So it can be a little difficult to find what you're looking for unless your subject has a unique, or at least unusual, last name.

Yet Grantor/Grantee records are good sources of information, such as addresses, full names, occupations, personal finance information, birth dates, and a lot more. The records are filed by year and by name, not a great source to approach through the mail, but workable if you can make a personal appearance at the desk and have a little patience.

## CIVIL AND CRIMINAL COURTS

### LOCAL COURT RECORDS

The typical local court lineup will include superior, municipal, small claims, probate, traffic, and divorce. Some jurisdictions have courts with other specific titles and functions: criminal and orphans, for example. Court records contain volumes of personal information about plaintiffs, defendants, witnesses, prosecutors, defenders. And every word of testimony and comment spoken in open court is taken down by a court reporter equipped with a special shorthand typing machine. A transcript of that record is available for a fee, which isn't cheap. The price can range up to $5.00 per page.

Most counties have a clerk of courts, or an equivalent, who is familiar with the workings of all the courts. Inquiries directed to that office should bring you the particulars about any court case in which your subject may be involved. Ask for a fee schedule when you write to inquire about your subject's court case.

These records are normally available to the public with the following exceptions:

- criminal cases in which the accused is a minor (under sixteen),
- adoption records,
- criminal cases involving a witness protected by the government,
- an order from the judge sealing the records under certain circumstances, as when confidential evidence is presented.

Confidential documents such as bank records, credit information, and other financial background data aren't normally available to the public. But you can get your hands on them through a *subpoena duces tecum* (court order) if they are critical to you for legal reasons.

Court records are usually filed in the names of plaintiff/defendant. You might find they are on microfilm, which you can review in a special booth. But they can't be removed from the records room. And divorce case records are only available by correspondence or by telephone.

It is no secret that the American system of justice is faulty. Nor is it a secret that ours is by far the best system available anywhere there is a verbal language.

Nowhere else on earth can ordinary citizens like you and me expect and receive the fairness that will ultimately be yours if you persevere in your demand for satisfaction from the courts.

## COURT FUNCTION BRIEFS

### Superior Court

This court usually handles felony/criminal cases and civil matters involving sums of $10,000 or more.

A *felony* is any crime from robbery to murder for which the defendant is tried by a prosecutor (district attorney) representing The People of the community and that is punishable by imprisonment of from more than a year up to the death penalty.

*Civil* cases deal with disputes between individuals over money, property, child custody, and so on. The People are not involved here unless the plaintiff is charging a unit of government with neglect—say, in a case involving an inoperative red light that resulted in a traffic fatality. Then a town or city solicitor will represent The People versus the plaintiff. Normally it's citizen against citizen, one (the plaintiff) charging the other (the defendant) with somehow violating the plaintiff's rights. The "sentence" for losers of civil suits is termed *damages*, and is usually awarded by the judge or jury in terms of dollars. In custody cases, loss or denial of custody may result.

### Municipal Court

The "Muni" deals with criminal misdemeanors, violations that upon conviction call for imprisonment of one year or less—minor theft, battery, drunk driving, and trespass, for example. Again it is The People versus the defendant, with the district attorney handling the prosecution. And—as you've heard so many times on TV as a cop "Mirandizes" arrested suspects—a defendant has the right to court-appointed counsel drawn from a pool of attorneys called public defenders who are on the public payroll. They represent those accuseds who can't afford to hire a private lawyer; this applies regardless of which court is hearing the criminal case.

Civil cases tried by the Muni deal with sums smaller than the lower limit set by the superior court (usually under $10,000).

## Small Claims Court

As its name implies, there is no crime presented here. It's citizen versus citizen. The dollar limit of this court varies from state to state, but sums involved tend to range from a few hundred dollars to about $10,000. An oddity about this court is that the defendant can bring a lawyer but, as a rule, the plaintiff must appear without one.

## Probate Records

These present the details of a decedent's estate, along with the names and addresses of those provided for in the will and the names of the executor, administrator appointed, lawyers involved, surviving and predeceased children, and the deceased's parents. You will learn also the final disposition of the will and the name of the presiding court officer.

These files are especially helpful in tracing birth parents or family roots, and they're yours—for a fee, of course—so if you write for a fee schedule first, you can include your payment along with your request for information.

## Traffic Court

These records are forwarded to the state by city, county, and state officials where they are filed at the Driver's License Bureau. You might try to get to them before they leave town, and they're available for a certain period after the violation. But you'd be better off to wait until they're safely at the capital, and a matter of record there. Local authorities—once you've been cited for a driving infraction—hesitate to monkey with documents destined for your all-important driving record.

## Divorce Court

The divorce or annulment decree usually gives the date and place of marriage, names of any children (excepting relinquished children but including adoptive ones), the couple's most recent address, the names of those persons from whom divorced earlier, a description and estimation of value of jointly held property. Records of the case are held at the courthouse where the decree was handed down, with copies sent to the state's vital statistics office. They are available by mail upon payment of a fee, which you should ask about before sending your request for copies. Give whatever of the following facts you know when filing your request.

- Full name and nickname(s) of husband and wife
- Present home address for each
- Other most recent addresses
- Ages at time of divorce or annulment
- Date and place of divorce or annulment
- Type of final decree
- Purpose for which copy is needed
- Relationship with the couple in question

(See Appendix 8 for addresses to which requests can be sent.)

## Grand Jury

The grand jury is a "court" in the sense that it decides whether or not a case will be prosecuted. It is convened periodically for the purpose of reviewing evidence against persons who are suspected of violating criminal law. Grand jurists can number twenty-three or more (the courtroom jury of twelve or fewer is called a *petit* jury), but only a quorum (majority) need be present to hear the evidence brought by the district attorney. The jurors decide, based on the evidence, whether or not to prosecute the person. A legal quirk prevents the defendant's being represented by counsel, nor is the defendant present to hear the evidence against him or her. If the grand jury finds evidence sufficient to warrant a trial, it hands down an *indictment.*

---

LETTER TO VOTER REGISTRATION

Director, Voter Registration
Courthouse
(Name and zip of county seat)

Regarding: Charles Thomas Ingham
            Date of Birth: 9/14/29
            In Louisville, Kentucky

(Date)

Dear Director:
    I realize that voter registration information, in spite of being public record, is not normally available through correspondence. I'm asking for an exception because of the distance and expense involved in my coming to you personally.
    Would you please use the enclosed self-addressed stamped postcard to notify me (1) if Mr. Ingham is registered at your office, and (2) if he is, the cost of a transcript of his registration information? My need for Mr. Ingham's address regards an urgent family matter
    I have enclosed a statistical profile of Mr. Ingham and a photocopy of my Alabama driver's license for my personal identification.
    Thank you for whatever you can do.

Peter J. Anderson (your signature)

Peter J. Anderson (typed or printed)
107 Maple Terrace
Montgomery, AL 36104
205-555-1234

Encl. (2)

---

## VOTER REGISTRATION

A couple of "ifs," here, regarding voter registration.

If you know in what county in what state your subject might be living, and if he or she is a registered voter, you may be able to retrieve considerable information. *Public* information. But . . .

It seems that the people who run voter registration offices by and large prefer that you make your inquiries in person. So your mailed kit may not work here.

But send it along anyway, with modifications. It might fall into the hands of someone who will understand your problem and give you the same information you'd get if you were to ask for it in person.

Response might include your subject's address, occupation, political party, probably the date and place of birth, and maybe even a Social Security number.

Look up the county seat in your atlas or at the library. Send a letter requesting any information that would be available to you if you were to present yourself in person. (See page 38 for a sample letter.) Enclose a stamped, self-addressed postcard along with a photocopy of your driver's license, which is probably the identification you would be asked for if you appeared at the counter.

## PUBLIC WELFARE

Don't expect an eager response to inquiries made of the welfare people. They deal with a mishmash of federal, state, and county administrations, which would put a crimp in any bureaucrat's public service enthusiasm. In any event—and maybe as a consequence of these tangled authorities and cross-filing—the law prescribes that the names of welfare recipients are *not* public information.

The above obviously offers initial discouragement, but maybe a commonsense appeal to an individual administrator might work.

Let's pretend your subject is a young, panicky, unwed mother-to-be. Or just an ordinary, discontented-with-family-life adolescent. They're both prime runaway prospects. And when the little money they might have runs out and they haven't got the will to come home, they just might turn up on the welfare rolls wherever they happen to land.

---

### LETTER OF INQUIRY TO WELFARE

Administrator
Public Services/Welfare Department
(County seat, state, and zip)

Regarding: Sarah Carol Goodman
         Date of birth: 2/3/73
         (Personal data encl.)

(Date)

Dear Administrator:
    I have reason to suspect my daughter (or whomever) named above is on your welfare rolls. We at home feel quite capable of dealing with and correcting the conditions that brought her to you.
    If you will supply us with her address (and phone number?), we will see that she is taken care of with generous affection. You can then replace her on your rolls with someone far more in need of care than she is.
    We would appreciate your cooperation. Enclosed is a stamped, self-addressed postcard. Call us collect, or better yet, have her call.
    Thank you for acting as promptly as your busy schedule allows.

Your name (signature)

Your name (typed or printed)
Your complete address

Encl. A photocopy of my driver's license, Sarah's picture and statistics, and a stamped, self-addressed postcard.
My telephone number is 123-555-1234.

---

If you have indications that your subject is in a certain area, identify the county seat from an atlas or a librarian. Write to the agency, using the sample letter on page 40 as a guide. Enclose a Profile and a photocopy of your driver's license, or that of someone at the same address. Include your stamped, forwarding postcard, this time addressed to yourself, along with your *telephone number* and a request that you be called *collect* in case the agency prefers calling rather than postcarding. With any luck, there will be confirming conversations or correspondence.

## WORKMEN'S COMPENSATION

In most states, all working people are covered by some form of compensation against loss of income through on-the-job injury. In a few states, including California, even legal aliens are so insured.

Although these records contain certain normally confidential medical information and background data, they are available to the public, usually at the county level.

They might provide you with the information you need about your subject. For more information, call the state's Workmen's Compensation Board. The number is in your phone book.

## LICENSES AND PERMITS

Today you need a license or permit to do almost *anything* commercial. In California (if anything has happened anywhere, it has happened in California) even door-to-door salesmen require a permit; in that regard, selling Girl Scout cookies without permits was an issue a couple of years ago in the Golden State. True. Some states require permits for spaces at flea markets and swap meets. Certainly roofers, plumbers, general contractors—almost anybody offering an entrepreneurial service to the public today—are, and should be, licensed.

Such documenting requires a great deal of exposure of information on the part of the licensee/permittee. These particulars are public information at the county courthouse.

# STATE RECORDS

## DRIVER/VEHICLE RECORDS

### DRIVER'S LICENSE RECORDS

American Express to the contrary, the driver's license is the one document *nobody* should leave home without. It has become the single most important piece of personal identification to be found in anyone's purse or wallet—the majority's link with respectability and the establishment. Visa and MasterCard may help a little, but a driver's license is much more reassuring to the law when a confrontation occurs. This is why many people in large cities have driver's licenses although they don't own a car; parking and liability insurance can be prohibitively expensive, but they can't function as viable citizens without a license as part of their ID. Thus, the "vehicle operator's license" becomes a prime source of personal information for the searcher.

Remember this major requirement in asking for driver's license information: *you must have the subject's date of birth and*

*correctly spelled full name.* And even these specifics won't guar-
antee a positive response from some states. The following
pages show the exceptions, which really aren't hard to deal
with.

It might save you time and aggravation if you were to
write simultaneously to the driver's license bureaus in every
state in which it seems logical that your subject would be driving: the one in which he or she was born; the state in which he
or she was originally licensed; the one you think he or she is
living in now; the one in which he or she vacationed or trav-
eled extensively.

The response may come in the form of a computer printout with your subject's driving record, showing accidents,
suspensions, revocations, date issued and expired, and so on.
Some states even provide a photograph.

Here are the states from which you can request a driving
record with only the full, correctly spelled name and department driver license number (DDL#), or just the name and date
of birth.

| | |
|---|---|
| Alaska | New Hampshire |
| Arizona | New Jersey |
| Colorado | New Mexico |
| Connecticut | North Carolina |
| Delaware | North Dakota |
| District of Columbia | Ohio |
| Florida | Oklahoma |
| Indiana (Soc. Sec. no. is DDL#) | Oregon |
| Kentucky | Rhode Island |
| Louisiana | South Carolina |
| Maine | South Dakota |
| Maryland | Tennessee |
| Michigan | Texas |
| Minnesota | Utah |
| Mississippi (Soc. Sec. no. is DDL#) | Vermont |
| Montana | Virginia |
| Nebraska | West Virginia |
| Nevada | Wisconsin |
| | Wyoming |

These states require the subject's written permission:

| | |
|---|---|
| Arkansas | Hawaii (Soc. Sec. no. is DDL#) |
| California | Kansas |
| Georgia | Pennsylvania |

These states require the subject's driver's license number:

| | |
|---|---|
| Alabama | Iowa |
| Idaho | Missouri |
| Illinois | |

The state of New York requires a DDL#, or the name and date of birth, and the subject's last known New York address.

Only two states, Massachusetts and Washington, will not supply driver's license information.

In some states, the driver's license number is area coded. So when you write for records information, ask for a code interpretation if one is needed. This could reveal the county or city where the license was issued, which would be helpful if it turns out that the subject is not living at the address shown on the driver's license. You could then concentrate your search in the area indicated by the code interpretation.

See Appendix 7 for addresses of all fifty-one driving record agencies.

### MOTOR VEHICLE RECORDS

Depending on the state's regulations, a check of the vehicle tag number will provide you with a computer printout of portions of or all of the following information: the name and address of the person to whom the tag was issued, all the vehicle's identification numbers, the owner's insurance company, date of registration and expiration, lien holder, year, model, and color.

You might even receive the owner's date of birth, which would be valuable in pursuing other information sources, including a driver's license, because most states require a date of birth (and the full, correctly spelled name). See Appendix 7 for the state departments of motor vehicle records.

### NATIONAL DRIVER'S REGISTRATION SERVICE

Another helpful information source is the National Driver's Registration Service. It was established to help law-enforcement agencies and insurance companies trace individuals with a suspended or revoked driver's license in one state who apply for a driver's license in another state. More details are available from: National Driver's Registration Service, U.S. Department of Commerce, 1717 H Street, Washington, DC 20510.

## CORPORATE RECORDS

Corporations active in the United States, whether American or foreign, must register in one of the fifty states. Their records are under control of the secretary of state, as a rule. They contain the date of the corporate filing, the name of the corporation, the names of the directors and officers, the headquarters location, the corporation's type of business, and other specifics.

There are two types of corporations—public and private: *private* means without stock on the market. The stock is privately held by an individual or by an individual and members of his official family (or board of directors).

The *public* corporation's stock is bought and sold in a recognized public market, and the records show the names of the officers and substantial stockholders. A publicly traded corporation must be listed with the Securities and Exchange Commission (SEC), a federal agency. The SEC allows the government to oversee all of the corporations' stock-trading activities and requires every such firm to submit financial details of major stockholders' financial activities.

In addition, every corporation must publish an annual report for the SEC and for its stockholders. The annual report contains details of the company's financial activities for the preceding year.

For more information on any corporation or any of its official personnel, write the secretary of state in the state where the corporation has its headquarters.

## STATE POLICE/HIGHWAY PATROL

Police at the state level take on different functions in different states. California is typical of most.

The *state police* there provide protection and security for government dignitaries and officials. They investigate crimes against the state, provide bailiffs for state-level courts, serve warrants for the State Franchise Tax Board. The state's *highway patrol* handles interstate and state highway traffic and accidents, almost its sole functions. (I can't imagine their being able to find time for anything else.)

Records of state law enforcement in most states are kept at the state capital, and most of them are public information. But whatever state your subject lives in, it's best to write first to learn if what you want is a public document and to ask what the fee is for a copy of it.

Some states provide car accident reports only to those involved and to the insurance companies. The reports are excellent sources of information and include names and addresses of victims and witnesses, vehicle license numbers, next of kin, and so on.

Vehicle accident records, when there are no state laws to prevent their being public information, can usually be found at the state capitol building, the local state office building, or the county courthouse.

## STATE TAX BOARDS

This agency, however it's known, is responsible for administering and collecting personal income tax and bank and corporation taxes. It has a clone in almost every state government, with larger or smaller tax-collecting responsibilities. You can expect little information from these folks, no matter what state you're in. But they'll be glad to explain what they can and can't talk about. Write to the state tax office in your subject's state capital.

# PERMITS AND LICENSES

### BOARD OF LICENSING

If your subject is in a *profession* whose practice demands a license—from hairdressing to brain surgery—his or her name is on file with the State Board of Licensing at the capital of the state in which he or she is operating.

You can write to the board and get such information from a professional license as correct spelling of name, middle name, address, telephone number, and other personal data.

### BOAT AND FISHING LICENSES

In a majority of states, the motor vehicle department handles the licensing of both commercial and pleasure boats operating on inland waters. These agencies also normally license the commercial and recreational fishing vessels operating out of the Great Lakes and seaboard states.

The fisherman himself is generally licensed by an agency titled Fish and Game Commission, or something similar. And in some states the licensing of both boat and fishermen is done by this one agency. But you're after records here.

If you address the Fish and Game Commission at the state capital, your inquiry is certain to reach competent people eager to see your questions answered.

## STATE SALES TAX BOARD

This agency or one like it administers and collects the state sales tax wherever there is such a law. Only products are taxed, not services. All businesses involved in the retail sale of products must file such a return.

These returns are available to you as a general rule, and they provide vital statistics regarding the business, including bank account numbers and estimated gross sales.

Some states have similar filings for businesses operating

as *wholesalers*. The wholesaler is not normally taxed for sales purposes but merely as an indicator of who is selling to whom and for statistics such as total sales volume.

Contact the sales tax board at the capital of the state your subject is living in.

# FEDERAL RECORDS

FREEDOM OF INFORMATION AND PRIVACY ACTS

It is now our right to demand and get at the federal level, information that affects us as individuals and as citizens, a right provided by the Freedom of Information Act. (Some *states* have passed their own FOIA.) All we have to provide is just cause for opening the records.

But until 1969 many of the methods that I'm suggesting in this book for acquiring even declassified information from the federal and state governments would have been denied or ignored.

This all changed when Public Law 5 USC 552A—known as the Freedom of Information Act—went into effect on July 4, 1967, after being signed into law by President Lyndon Johnson. It has had a powerful and controversial impact on the status of secrecy regarding federal government records. It was used by courts and the media to expose the affairs of Watergate. It opened the files on gifts from foreign officials to President Nixon and caused the Defense Department to release details of the My Lai massacre in Vietnam. Recent government efforts to investigate itself—Waco, Ruby Ridge, CIA intelligence aberrations, etc.—would never have been done in public view had it not been for the FOIA. As you can see, the act is quite powerful.

The Privacy Act of 1974, 5 USC 552B, was intended to and does limit the federal government's authority in collecting and using information on individuals. At the same time, it somewhat broadens and supplements USC 552A. In any event, don't hesitate to cite either or both acts in requesting information from federal agencies. And be sure to read both acts at your library.

See below for a guide to writing your own memo on these acts, and include a copy with your letter of inquiry to let the officials know you know they must comply.

---

### USING THE ACTS

To Those Concerned:
    Pursuant to the Freedom of Information Act (5 USC 552A) and the Privacy Act (5 USC 552B), I request copies of all records regarding myself maintained by your office.
    The following descriptive data are supplied solely for the purpose of identifying myself, and are not to be released to any other agency or individual:
    Name:
    Date of birth:
    Address:
    Passport:
    Social Security number:
    Hereon is notarization attesting to my true identity.
    Please send a summary of the information requested, along with copies of my records.
Very truly yours,

(Your signature)

(Have a notary public attest your identification in this space.)

---

The Freedom of Information Act (5 USC 552A) and the Privacy Act (5 USC 552B) have been the source of some misunderstanding and misapplication. You're referred to an excellent pamphlet produced by the American Civil Liberties Union (ACLU), *A Step-by-Step Guide to The Freedom of Information Act*. Write to the ACLU, 132 W. 43rd Street, New York, NY 10036

## SOCIAL SECURITY

Dealing with federal agencies is not always a simple matter. But here and there you encounter open doors, minds, and individuals with a desire to be helpful.

Such is the case with the Social Security Administration. It probably has the largest roster of names of any agency in the United States government, since most American adults have a Social Security number and a file of some kind. The chances are good that your adult subject, whether he or she is employed or has retired, will have a reasonably recent address available—either a home address, the address of an employer, or both. If the subject is receiving Social Security insurance checks, or compensation for any reason, the presence of a home address is very likely.

In any case, Social Security has formulated a policy of allowing mail contact "for humanitarian purposes" with anyone on its roster whose address is available. Simply send your Inquiry Kit, including your subject's Profile, and your forwarding postcard will be sent to the subject. Allow for delay, as the search for a file will take time. And try to give the message on your postcard as much "humanitarian" feel as you can without fabricating.

If you do not include a Profile or a Social Security number, be certain that you *do* provide your subject's full name, and date and place of birth.

See page 52 for an idea of how to formulate your letter to the Social Security people.

---

### LETTER OF INQUIRY TO THE
### SOCIAL SECURITY ADMINISTRATION

Director, Locator Service
Social Security Administration
6401 Security Boulevard
Baltimore, MD 12135

Regarding: Charles Thomas Ingham
            Date of Birth: 1/14/54
            Louisville, Kentucky

(Date)

Dear Director:

    It would be a great help to the family of the above-named person if you were able to forward the enclosed stamped, unaddressed postcard to him at the latest address you have available. It contains a message requesting that he get in touch with his family. If you are unable to supply a reasonably recent address, would you kindly see that the postcard is mailed back to me.

    Thank you. I have enclosed a statistical profile of Mr. Ingham as well as a photocopy of my Alabama driver's license for my own identification, along with the postcard.

Peter J. Anderson (your signature)

Peter J. Anderson (typed or printed)
107 Maple Terrace
Montgomery, AL 36104
Encls: 3

---

### THE SOCIAL SECURITY NUMBER

The Social Security number is important in your search, and you should do all you can to get it. The best place to look for it is in credit reports, which we will discuss later.

    A Social Security number provides verification of identification. Everything from credit reports to business license applications contain or use Social Security numbers. For example, my insurance company uses mine as an index for their files—I don't know why or how. But the Social Security number has pretty much become the American serial number. And the administration wants to have a number issued to every newborn.

In addition to maintaining records on virtually every American, the Social Security Administration keeps track of millions of foreigners who work in this country or who once worked in this country and have since retired to live outside the United States.

Except for a few numbers issued in the mid-1970s to military recruits, all Social Security numbers contain nine digits. Those military Social Security numbers contained ten digits beginning with zero. There are very few of them.

The first three numerals are known as "area numbers" because they indicate from which state the subject applied for a number. This may be a clue to where to begin your search—maybe with a check of driver's license records for that state, which could bring you an address if the subject still lives there.

For an index of Social Security numbers, see Appendix 1. Remember, Social Security records are confidential and not available for public or even law-enforcement review.

## THE U.S. POSTAL SERVICE

Surprisingly, if you know the address where your subject was living at some time during the past six months, you can get the address to which he or she moved. Simply send the last known address, along with a dollar, Attention Postmaster, to the appropriate post office, and a forwarding address will be provided. If the subject has been gone for more than six months, however, you're out of luck. That is as long as the postmaster holds a forwarding address.

If you want to save a half-dollar—and who doesn't?—mail a letter to the old address with DO NOT FORWARD—ADDRESS CORRECTION REQUESTED on the envelope. The postmaster will deliver the new address back to you and charge you 50¢.

There are other postal service features you should be aware of, such as certified mail, registered mail, and express mail. Return receipts are available for each. *Certified* mail provides a receipt as proof of mailing. Request a return receipt to show when and to whom the mail was delivered. You might want to use maximum-security *registered* mail when you are sending a valuable document (you determine the value), and you can restrict delivery to the person addressed. Your mailing will be returned to you if the person has moved. *Express* mail guarantees priority treatment and minimum delays. All of these services demand only relatively modest fees.

Use *General Delivery* as a return address if you prefer not to reveal your home address to whomever you are writing. There is no charge for this. General Delivery mail is held at the post office for thirty days for your pickup. At the end of that time, it is returned to the sender.

Individuals and businessmen using a post office box are required by law to file a valid street address. Such an address for a business will be furnished to you upon request. An individual's address, however, is not available without a court order (*subpoena duces tecum*).

If you think you have located your subject at a particular address, you can ask the postal carrier to help you determine whether the addressee is the subject you're looking for. It's worth the effort to try for identification this way, but don't be surprised if the carrier doesn't cooperate. Be ready with a humanitarian motive for your search if you want that postman's help.

## U.S. DISTRICT COURT

The jurisdiction of the U.S. District Court is based on federal statutes and the U.S. Constitution. It handles both civil and criminal matters.

Any cases decided here can be appealed to the federal appellate level and from there to the top: the U.S. Supreme Court.

The source for retrieving information about principals in any cases decided in the appellate court or Supreme Court is here at the U.S. District Court, neatly indexed and ready to be examined at the federal courthouse. Or write to the clerk of the U.S. District Court in the federal courthouse nearest your center of interest.

## BANKRUPTCY COURT

Lots of people and companies in financial trouble file for the protection of bankruptcy, which is handled in a federal bankruptcy court. If you have reason to believe your subject has ever filed for bankruptcy, you can start your search of records in the federal court clerk's office nearest the site of the bankruptcy. You will find there all the details you will need for ascertaining a full, correctly spelled name; age, address, and net worth at the time of filing; and name and kind of business. Also, spouse's name, age, and occupation, if any. All of this is public information, and can be had by mail.

## MILITARY LOCATORS

Don't overlook the possibility that your subject may be in the armed forces. If so, there should be no problem finding him or her. All the services now have computerized locator sections. Send a Profile with the following details highlighted:

- Name
- Service serial number (may be a Social Security number)
- Last known address
- Date of birth
- Latest rank

**U.S. ARMY**
Worldwide Locator Service
U.S. Army Personnel Service Support Center
Fort Benjamin Harrison, IN 46249
317-542-4211

**U.S. AIR FORCE**
Air Force Military Personnel Center
Worldwide Locator
Randolph AFB
San Antonio, TX 78150
210-652-5775

**U.S. NAVY**
Navy Locator Service
Bureau of Navy Personnel
Drawer #2
Washington, DC 20370
703-614-5011

**U.S. MARINE CORPS**
Marine Corps Headquarters
Locator Service
2000-8 Elliot Rd. Ste. 20
Quantico, VA 22134-5030
703-784-3942

**U.S. COAST GUARD**
Coast Guard Locator Service
Room 4502 (enlisted)
Room 420B (commissioned)
2100 2nd Street, SW
Washington, DC 20593
202-267-1340

**RETIRED MILITARY AND CIVIL SERVICE PERSONNEL**
The Office of Personnel Management
1900 E. Street, SW
Washington, DC 20415

## MILITARY RECORDS

For all military records dating back to World War I, send your Inquiry Kit, which should include the subject's military serial number, discharge date and rank, service branch, combat theater, and a statement that you are a relative, to, General Services Administration, National Personnel Records Center, 9700 Page Boulevard, St. Louis, MO 83232.

## U.S. MARSHAL

Is there reason to think your subject might be a fugitive from the law? If so, contact your nearest U.S. Marshal's office and inquire about your subject. Give the marshal only the full name and date of birth, no more. Ask that you be given whatever information might be available from the National Crime Information Center (NCIC) in Washington. It's against the law for the marshal to give such information to the public. He knows that. He doesn't know that you know. So be aware that if he comes back to you needing more information on your subject before he can help you, that's a signal your subject may be among the wanted.

There's a U.S. Marshal in the white pages of your phone book under "U.S. Government." His main headquarters is at 600 Army-Navy Drive, Arlington, VA 22202 (telephone: 800-336-0102).

## THE NATIONAL CRIME INFORMATION CENTER (NCIC)

NCIC is a computerized system for storing and retrieving crime information. It's located in Washington, D.C.

NCIC is maintained by the Federal Bureau of Investigation and contains millions of names, hundreds of thousands of stolen property descriptions, and the criminal histories of thousands of lawbreakers.

Linked to every law-enforcement agency of any size in the United States, the system has been instrumental in the apprehension of many hundreds of criminals and the recovery of millions of dollars in stolen property over the few years it has been in service.

For further information on NCIC and how it can be helpful to you, ask to speak to the public information officer at your police or sheriff's department, or to the U.S. Marshal in your region.

Be aware that you can ask your local law-enforcement agency to enter the name of any missing child into NCIC. If refused, you can go straight to the nearest FBI office with your request.

To qualify entry of a name into the NCIC system, a juvenile must be under legal age in that state and an adult must be physically or mentally disabled, or foul play or suspicious circumstances must be involved in his or her disappearance. If police or the sheriff refuse to enter the qualified missing person's name, go to the nearest FBI office with your report. The law says it *must* be accepted and entered into NCIC.

## INTERNAL REVENUE SERVICE (IRS)

The IRS is a government agency with especially tight security. At one time it was possible to request the latest date of a tax return for an individual or business that would include the address to which the return was sent. This service has been retracted.

## INTERSTATE COMMERCE COMMISSION (ICC)

ICC records are useful if your subject happens to be a trucker and hauls goods across state lines. They are public information and are available even by phone. This agency regulates and licenses every form of interstate transport, and it's listed in the white pages under "U.S. Government."

## ALCOHOL, TOBACCO, AND FIREARMS (ATF)

The stamps you see on alcohol and cigarette containers are the doings of the ATF, along with the issuing of licenses to sell firearms. Anything that makes you feel good or that tastes good or makes a loud noise—the ATF has a record to go with it. And that information is public.

## THE CENSUS TAKERS

If you have a *legal* reason for your search, the Census Bureau may be able to help you with a reasonably current address. There is a form you must send for: No. 106–11. Write: The Bureau of the Census, Pittsburg, KS 66762. There is a fee. Ask about it when you send for your inquiry form.

## GOVERNMENT PRINTING OFFICE (GPO)

The GPO is the most prolific book publisher on earth. Every year it prints hundreds of thousands of books and pamphlets on every subject imaginable. The GPO catalog of titles is like an itemization of man's total knowledge. But for our purposes, one of its most significant publications is—of all titles—*The Information Book*. It contains the names, addresses, and phone numbers of every government office—and its principal occupant—in the United States today. Write: Superintendent of Documents, U.S. Government Printing Office, Department 33, Washington, DC 20402; or telephone 202-275-2481.

## SELECTIVE SERVICE

A number of changes in location and procedure have taken place in the Selective Service in recent years, and apparently are still going on.

So for up-to-date information regarding your own or your subject's civilian/military status probably your best bet is to contact the Selective Service National Headquarters, 1515 Wilson Boulevard, Ste. 400, Arlington, VA 22209; or call 703-235-2555.

There was a time when Selective Service could release only the City and State locations of anyone inquired about. Information these days is a little more detailed.

Simply supply full name, address, and if there is one available to you, your telephone number.

## FEDERAL OFFICE OF CHILD SUPPORT ENFORCEMENT (FOCSE)

The last nose count of missing parents of dependent children—most of them fathers, by far—comes to almost 9 million in the United States. The statisticians who project such figures calculate that by the 1990s more than 50 percent of the kids in the United States will spend their entire childhood with just one parent—in most cases the mother, as you would expect.

The U.S. Department of Health and Human Services has a division called the Federal Office of Child Support Enforcement that is attempting to do something about the enormous tab—$25 billion or so annually—that aid to dependent children is costing the taxpayers. About 60 percent of the $11 billion in support owed in 1992 has been collected from absentee fathers by intercepting tax-refund checks, mandatory income withholding, liens against property and securities, and other means. And the system is getting tougher and working better all the time.

Every state has its own version of FOCSE, usually administered by the counties through either the district attorney's investigation division or by the welfare department.

If the father of your dependent children is among the missing, whether or not you want to locate him, the place to start a collection action is right there in your own county. Call your local district attorney's office or the county welfare department for information regarding FOCSE.

## FEDERAL AVIATION AUTHORITY (FAA)

The licensing of planes and pilots is a function of the Federal Aviation Authority. There are no other agencies involved. If your subject is a flyer and you wish to gain information about his or her location, approach the FAA as you would any state driver's licensing bureau. Write Federal Aviation Authority, Mike Monroney Aero Center, 6500 South MacArthur Drive, P.O. Box 25082, Oklahoma City, OK 73125. To the left of the address on the envelope, write this code: AAC260. The telephone number is 405-954-3001.

You might be interested to learn that although all airplane licenses are the responsibility of an agency of the U.S. government, the plane—like a car in any state—is considered unsecured property and is subject to all state legislation dealing with division of property, inheritance, and legal seizure.

Your subject may subscribe to one of the aviation magazines. Check your library for addresses, and send each circulation manager an Inquiry Kit.

## PRISON RECORDS

It may be depressing to think of your subject as a felon, a wife-beater, or a drunk driver. But, life being what it is, it's conceivable that your subject is behind bars.

Four prison systems exist in the United States: *federal*—for persons convicted of violating federal laws; *state*—reserved for those in violation of state laws and local statutes; *county*—

where those awaiting trial, being tried, paying dues for non-support, serving weekends for drunk driving or an overload of traffic violations all may be found; and *city jail*—overnight drunk tank, spitting-in-public-type lawbreakers.

The addresses for all of these systems, of course, are available at your library. Use your Inquiry Kit. Make your message to your subject on the forwarding postcard as simple and un-provocative as possible—even innocuous. You don't want to cause your subject, or yourself, any more trouble than he or she has got right now. Asking a person to contact you is OK. Maybe that's all you need to say. Prison inmates have writing privileges, although their mail might be censored. (All their phone calls must be made collect.)

Just don't bank on getting an answer, given the circumstances—and depending on what kind of thought your message imparts—your subject may choose not to respond.

# MISCELLANEOUS

## EDUCATIONAL CHANNELS

Let's make a case history of this usually rewarding way of locating someone through educational channels.

**Subject:**   A fondly remembered classmate.
**Subject's full name:**   Mary L. Dwyer, nickname "Boots"
**Description:**   Caucasian, 5'2", blond, blue eyes, dimples, glasses, slender (at last contact)
**Last known address:**   812 Grandview Avenue, Pittsburgh, Pennsylvania 15211
**Graduated:**   South Hills High School, June 1985, Academic
**Whereabouts:**   Unknown
**Reason for search:**   Affection, desire for re-contact

Using a commonsense rule, you begin your search with the last known facts about your subject: graduate of South Hills High School, June 1985, the month you joined the Marines and she moved away from the city. Ten years later, you're on the East Coast and have completely lost track of classmates who may know where she is. You send your Inquiry Kit to South Hills High, including the Profile, incomplete as it is.

Your postcard is returned to you, noting that the school has no current address for Mary. But the counselor's records indicate Mary's intention of attending Polk Community College in Atlanta, Georgia, where her family was planning to resettle.

You get Polk's address from the library and send your Inquiry Kit to the registrar's office. No current address, says the returned postcard; try the alumni association for all Georgia community colleges, located in Augusta.

The alumni association reports that Mary took a photojournalism degree from Polk, married Ralph Angelo, and is living at 567 Mariposa Street, San Francisco, California.

Your letter to Mariposa Street returns: NO LONGER AT THIS ADDRESS. And the request for a forwarding address is returned with your dollar: Mary's been gone from there for more than a year.

Dead end? Not quite. The reference librarian at the San Francisco main library, at your written request, checks the Haines Criss-Cross Directory for the present resident at 567 Mariposa Street. Whoever it is doesn't have a telephone, so the name isn't listed, and you can't just write to "Occupant" for the information you need. But the Criss-Cross shows also that Mary's former neighbor has a phone and a name with a special ring to it—Felicity Gallagher.

It's time to bring the chase to a close or forget it. You pick up the phone at 11:05 P.M. your time at Chevy Chase, Maryland. It's 8:05 P.M. Pacific time, with Felicity quietly watching television in San Francisco, just as you'd hoped. It takes a while to get this pleasant, talkative lady to the point. But finally you learn that she was not only friendly with Mary Dwyer but can report that Mary has no children and is no longer married to Angelo. Eighteen months ago Mary left to take a job at a Sacramento, California, newspaper. Which one? "A big one," says Felicity.

Well—"the big one" in Sacramento is *The Sacramento Bee*, big enough that the operator asks, "Which department?" Try photography.

There is a buzz, a click, and then the sweet voice you haven't heard for more than ten years is on the line.

"Photo lab, Mary."

## NEWSPAPER PERSONALS

Contact through newspaper personals may seem a bit far-fetched to some people. But there isn't a regularly published newspaper in the country whose classifieds don't include these private messages. And it's a way to go if you're limited to knowing only your subject's approximate whereabouts. That location should be within circulation reach of a substantial news periodical; find the publication nearest your target area by consulting *Ayer's Dictionary of Newspapers and Periodicals* or *R. R. Bowker's National Directory of Weekly Newspapers*. These references will also provide you with the line or word rate, so you can send along a certified check or money order and ensure that your message gets published on schedule.

I suggest you run your ad at least three times, and more if you can afford it. Your message should be brief, of course, if you're concerned about cost.

For example: "Urgent Della Benedict contact Tim. Write care of General Delivery, Rancho Mirage, CA 92270. Or telephone collect through information."

We're assuming Della knows who Tim is. He's using General Delivery to avoid publishing his home address, and the collect call through information for the same reason. Otherwise, the cranks can have a picnic harassing personals advertisers.

Now let's hope Della or some of her friends are avid personals readers.

## NEWSPAPER "MORGUES"

Newspapers keep an index of all back issues. This facility is called the morgue, because—as the saying goes—there's nothing deader than yesterday's newspaper. But you could be lucky and find a real "live" back issue.

The morgue is an indexed-by-name file, and is available to you for a modest fee. If your subject has been in the local news, the clerk can find—or show *you* how to find—the mention of his or her name (and probably an address, if it has appeared in a reasonably recent past issue). The subject needn't be a headline type. A name is more likely to be found in a wedding announcement, a legal message, a birth notice, a publicity release about a job promotion, and so on. Even an obituary could help resolve your search.

## CATALOG MAILING LISTS

Sears, Montgomery Ward, Spiegel's, TV Home Shopping, and other mail-order houses maintain updated customer addresses. That's their business. Here is a rather long shot that may not be all that long if you know your subject to be a "send-away-for" shopper, like millions of other Americans.

You might want to "shotgun" your inquiry by sending your kit to two or more mail-order retailers. If there is an obvious same-name problem, press your luck and send several forwarding postcards to each retailer, and try to give a general location for your subject. (The message on the postcard should indicate some kind of urgency in order to gain the interest and sympathy of the mail-order boss.)

The addresses of all the mail-order houses are available at your library. Direct your inquiry to: Supervisor, Customer Mailing Lists.

## UNIONS AND ASSOCIATIONS

There is an old joke about Americans and their organizational zeal: two Americans meet overseas for the first time, say hello, and shake hands; *three* Americans meet overseas for the first time, say hello—and elect officers.

If your subject has a craft, business, profession, or talent of some kind, he or she probably belongs to *something*. Dentists, lawyers, architects, railway workers, CPAs, carpenters, liquor store clerks, animal breeders, and fiddle players are *all* members of some guild or union. Even retired people have a union of sorts: The American Association of Retired Persons (AARP) is one of several such groups.

Or your subject may be just a person who likes companionship and the chance to do good with one of the many service groups: Masons, Moose, Elks, Rotary, Variety Club, and so on.

So your chances of locating an individual using this route are good if you know your subject's trade, profession, or fraternal/social leanings. Most of these organizations have an interstate apparatus with a national headquarters. Many of them issue "house organs," or newsletters, that publish pictures of members. You might get your subject's picture, or at least a brief inquiry about him or her, into one of these periodicals.

Of course, your library has all the information you need. Direct your kit to the national secretary at the organization's national headquarters.

## INSURANCE RECORDS

If you are familiar with the name of your subject's life, home, or auto insurance carrier, a door opens there. People tend to stay with the same insurance company for decades, regardless of relocation. Your reference librarian will help you with the address of the insurance company's headquarters, where the updated records of their insureds remain on file.

In your inquiry letter, make your reasons for wanting to contact the subject brief, clear, and appealing. Have a good, warm reason for wanting contact. Emphasize your understanding of the company's need to protect its insured's privacy and make it clear that your forwarding postcard won't violate that privacy.

## MAGAZINE SUBSCRIPTION LISTS

Is your subject a bird-watcher? A recipe collector? Does he or she breed Dalmatians, live on crossword puzzles, or carve little flowers out of peach stones? Is your subject an incurable jock, or a musician? A doctor, hairdresser, or—almost whatever?

Today there is a publication for every interest imaginable, from hang gliding to deciphering hieroglyphics. If you know your subject's obsession or profession, you may have something, because he or she no doubt subscribes to a magazine devoted to that interest.

All regularly published American magazines keep up-to-date subscriber lists. And as they're essentially in the public relations business, you can expect them to cooperate in the simple process of transferring an address from their computer to your forwarding postcard. This could well nourish their subscriber's goodwill and probably improve the likelihood of a subscription renewal.

Get the name and address of the appropriate magazines from issues on your supermarket rack or from *Ayer's Directory of Newspapers and Periodicals* at the library. Magazines are listed there under subject matter.

Your kit goes to the subscription supervisor. Try to give your subject a general location. Make your postcard message as urgent as possible without declaring a national emergency. If there are two or three magazines involved, send a kit to each.

## FYI—TELEPHONING FREE

There are inexpensive paperbacks in most bookstores useful in researching names, addresses, fax, and 800 numbers for hundreds of businesses and associations. Also included are U.S. Government agencies, the media, and dozens of service organizations.

One is the *National Directory of Addresses and Telephone Numbers* by Omnigraphics, Inc., consumer oriented, easy to use, lists toll-free numbers. A money saver if your search leaves you that way—contacting personnel offices, for inquiry about employees, for example. And it has many other applications for use in everyday life. You will be amazed at the number of organizations, civil and otherwise, you can contact by phone—free.

AT&T publishes two such books listing its own 800 numbers. One for consumers: *AT&T Toll-Free 800 Numbers; AT&T Toll-Free 800 Directory for Business*. Write AT&T, P.O. Box 44068, Jacksonville, FL 32203. Both are reasonably priced.

Oh, yes, 800-426-8686. You can use that to get prices.

## THE SALVATION ARMY

The "Army" was mentioned earlier in that tale about the Grand Central Terminal hero, one of thousands of accounts that have become legend about this remarkable organization. It has been in business since just after the Civil War. Among many other services, it provides shelter and free meal facilities in almost every medium-to-large city in the United States.

It even has a missing persons bureau.

If you suspect your subject is down on his or her luck and has needed to take advantage of the Salvation Army's hospitality, write to one of these geographically located Army headquarters, directing your letter to: Salvation Army Missing Persons Service.

**Eastern United States:** (Connecticut, Delaware, Maine, Massachusetts, New Hampshire, New Jersey, New York, Ohio, Pennsylvania, Rhode Island, Vermont) P.O. Box C335, W. Nyack, NY 10994-1739

**Central United States:** (Illinois, Indiana, Iowa, Kansas, Michigan, Minnesota, Missouri, Nebraska, North Dakota, South Dakota, Wisconsin) 10 W. Algonquin Road, Des Plaines, IL 60016

**Southern United States:** (Alabama, Arkansas, Florida, Georgia, Kentucky, Louisiana, Maryland, Mississippi, North Carolina, Oklahoma, South Carolina, Tennessee, Texas, Virginia, Washington, D.C., West Virginia) 1424 N.E. Expressway, Atlanta, GA 30329

**Western United States:** (Alaska, Arizona, California, Colorado, Hawaii, Idaho, Montana, Nevada, New Mexico, Oregon, Utah, Washington, Wyoming) 30840 Hawthorne Boulevard, Rancho Palos Verdes, CA 90274

Include all items in your Inquiry Kit—Profile, postcard, letter, and the most generous contribution you can come up with for this remarkable organization. Explain your genuine concern to the officer-in-charge very carefully. Have a positive reason for having your card forwarded if your subject is on the Salvation Army rolls. You could be dealing with your subject's pride here, so your reason should be a good one: "Let me help you" is usually enough.

## RELIGIOUS AFFILIATIONS

The possibility of contacting your subject through a religious affiliation might seem farfetched. But this is a good lead if he or she happens to be a regular tither in a prominent denomination. All denominations keep accurate records of donors and tithers; most have well-developed national headquarters and interstate communication networks.

You would need some notion of your subject's location. But given, say, the Grand Rapids, Michigan, area and the fact that he or she is a service-attending Baptist and a regular tither, you have the beginning of a lead. Your reference librarian will help you find the address for the national headquarters of any denomination.

Write and ask the *national secretary* (of whatever denomination) for help in getting your message to all its pastorates in the designated area, and request a thirty-word announcement by the ministers during services.

Send the following to the national secretary: an explanatory letter, your subject's Profile, and a modest donation for the reading of the announcement.

Some Sunday—or Saturday—your subject may be surprised to hear his or her name called from the pulpit to stay after services for a message from a "higher authority."

There could be something more than ordinary hope in trying to reach a loved or much-needed one in this way.

It's called *faith*. And I have faith that it can work. Try it.

### WHO'S WHO IN AMERICA

Might your subject be a person of considerable consequence financially, politically, professionally, artistically? A *Who's Who in America* personality?

There is such a registry for overachievers in all the above categories. Not every one of these people is a household name, but some are big in their fields—which can be as diversified in nature as you could ever imagine.

If you can see your subject as possibly being one of these exalted persons, you might as well check him or her out in the annually updated volume titled *Who's Who in America* at the library. All you have to know is the category in which he or she might be selected: science, government, literature, medicine, industry, and so on. The *Who's Who* entry's little bio is a business address, and a summation of the subject's contribution to the specialty.

If your subject is a business executive, he or she might be listed in a volume titled *The Business Guide to Corporate Executives*, also updated annually, listing the chief executives of nearly every corporation in America, with their addresses and phone numbers.

Then there is a massive tome called simply *The Red Book*. This lists not only the top echelon of every advertising agency by name but also gives a list of the agency's clients, the clients' advertising managers, and the amount of each client's advertising budget for the previous year. If you're searching for whoever is responsible for a television commercial you've grown to hate, this is the book for you.

## HOME UTILITIES

Water, gas, electric, and trash collection records are normally confidential, but information on them can be obtained with an appropriate pretext. Call customer service and say you have received a bill in your subject's name and will forward it to him if you can have his address.

## THE TELEPHONE COMPANY

Unlisted telephone numbers are given at the request of the subscriber. Amazingly, there is a fee for this nonprivilege. Why? Because the subscriber's name must be pulled from the computer before the directory goes to press, and he or she must pay for this service.

Anyway, this unlisted, unpublished number is confidential. The only way to get it legally is through a court order. The only way to get it *illegally*—for a fee—is to know someone inside the phone company who is into that kind of game. This is a private investigator tactic, but your subject's itemized bills for toll calls—both local and long-distance—are available for a price from unscrupulous telephone company employees. Or you could have a pal working there get it for you free. Surprisingly, it is not against the law to use information gained in this manner.

As an aside, I am involved in a number of sensitive investigations that require me to be in contact with confidential sources on a regular basis. I use pay phones frequently to maintain security. I code each one and note the pay phone's telephone number and code number in my address book. I then provide my sources with a code and phone number, and have them call me at that location: "Use Number Ten [the location code number] at three o'clock in the afternoon on Tuesday." My sources use the same technique.

It's not 100 percent safe, but it is unlikely that my conversations will be intercepted.

### TELEPHONE TAPS AND TRAPS

Before we go back to search methods, let me give you some tips on how to maintain telephone security.

If security is urgent and interception is even a remote possibility, never discuss your business on a regular or cellular phone. Both parties should use pay phones, and phone locations should be changed regularly. The newest telephone eavesdropping equipment is incredibly sensitive and intrusive, and illegal taps are prevalent throughout the United States. The perpetrators are technicians both inside and outside the phone companies, and it is virtually impossible for them to be detected and prosecuted. This is a violation within the jurisdiction of the FBI.

Telephone traps provide a "lock-in" on a phone being used to call a particular number, allowing the location of the caller to be determined. Law-enforcement agencies regularly use this device—with the phone company's help—when investigating extortion, kidnapping, and related crimes. They have not been above using phone company employees illegally in their investigations.

Electronic science has recently developed an inexpensive,

over-the-counter, portable trap. So if you don't want the person you're calling to have access to your phone number—and thus your address, with the help of the Criss-Cross Directory—use a pay phone to be safe.

## CHAMBERS OF COMMERCE

These organizations keep track of merchants, small businessmen, and professionals in their communities. They pretty much know what's going on around town, and who the new arrivals in town are, through welcome wagon activities and so on. The Chamber (which might be called the Board of Trade in some towns) is a pretty good source of information about people in towns.

Address your kit to: Chamber of Commerce, Attention: Secretary. It will go to the right people, even if the group's name is different.

## CLIPPING SERVICES

These firms read and clip just about every publication printed. They index by subject matter and name.

Bacons: 800-621-0561
Burrelle's Press: 800-777-8398
Lace: 800-528-8226

## BOOK SOURCES

*National Directory of Addresses and Telephone Numbers*
850 Third Avenue
New York, NY 10022

This book lists every 800 number, member of government (including the president; his office telephone number is 202-456-1414), bank, law firm, trade union, airline, and accountant.

*Confidential Information Sources, Public and Private*
Butterworth Inc.
10 Tower Office Park
Woburn, MA 01801

This book was written by John M. Carroll, published by Security World Publication Company. Its Library of Congress Catalog Card Number is 74-20177.

# COMPUTERS, CREDIT,
# AND CONSUMERS

American adults leave "paper trails" wherever they go. We live in a society that requires us to spread our identity over most of our environment. Our need for credit cards, checking accounts, doctors, insurance, subscriptions, shelter, utilities, taxes, and an ocean of services demolishes anonymity.

This chapter is directed for the most part to searchers for whom following a paper trail is the only recourse. It deals primarily with information sources designed around that most unsentimental of today's wonder devices, the computer, which provides information impartially to friendly and unfriendly hunter alike.

The advent of Internet and its wealth of all kinds of information is an exotic source for people-searchers use, since it requires special facilities and skills. There are other computer-driven ways of following a paper trail. But Internet first, FYI.

## INTERNET

You've heard of Internet. If not, you have been on a totally isolated sabbatical for the past five years. Or your communication senses are unfortunately not in working order.

Internet is the culmination of our civilization's gluttony for wanting to know everything about everything, even sometimes when what you might learn is of questionable value to you.

It spans the earth and links your computer with millions of others. It's a network that allows you to "browse" or "surf" through databases in search of your subject matter, and to bring up the material you need, instantaneously. One IBM model can be operated vocally! Obeys the spoken word.

Internet can be useful to you in your search, *if* it is not beyond your means. If you must start from scratch with the purchase of an up-to-date computer, you may be into a negative cost-effectiveness area. If you already have an up-to-date computer, you must know more than I do about Internet's capabilities. It's also likely you are already on-line.

But this incredible source of information could be available to you at larger public libraries, units made usable for beginners with simple "how to" in big-print instructions. Even I—all thumbs—was able to coax the terminal at Santa Monica's main library into bringing me the names of all the members of the California's public school teachers' union. This was from a database I picked at random to test its application to people-search.

Vendors of Internet, America Online, etc., services are available everywhere.

### OTHER INFORMATION SOURCES

There are many commercial computer facilities that can provide personal data on individuals for checking background information, employment histories, credit ratings, real estate property owned, and so on. In fact, for a fee these sources provide any kind of financial information on individuals or companies, from bankruptcy to their latest net-worth figure. Some facilities also provide searches of just about any public record that exists, including those for driver's licenses, real estate transactions, and death certificates.

I have listed a few of the more prominent facilities here. You will pay fees for services in each case, of course.

### BROOKLYN BUSINESS LIBRARY

If your subject is involved in a business of any kind, and specifics about that business or the names of firms become important, here is a regular cornucopia of information. The Brooklyn Business Library is one of the principal sources of reliable business information in the United States. Another of its attractions for the searcher is the *accessibility* of the data stored there.

If your need is for the name of a firm with whom your subject is associated, for example, or what have been the recent returns on stocks or bonds you know your subject holds, the Brooklyn (NY) Business Library is the source to turn to, and you can do it by telephone.

The number of the reference desk at the library is 718-722-3333.

### FEDERAL HELP

Aside from the remarkable family history records available through the Church of Jesus Christ of Latter-Day Saints (Mormon) are the National Archives files in Washington, D.C. A very accommodating reference staff there will furnish you with contemporary holdings information and for a modest sum will supply you with documents.

The number of the reference desk is 202-501-5400.

### NATIONAL DATA RESEARCH CENTER

National Data Research Center is a keeper and researcher of real estate records, a service of great value to you if you know that your subject is prone to dealing in real estate. According to its literature, NDRC is capable of researching in any of 300 counties and in 32 of the continental United States. All of these ownerships, by county, are updated yearly. Current address of the property owner is part of the information package, NDRC states.

## INFORMATION ON DEMAND

IOD's literature offers sophisticated fact collection and figures structured around finance, government, import/export, investment, legal, patent/trademark, and scientific/technical record searches. Names and addresses of principals involved in investment ventures are supplied. Write: Information on Demand, P.O. Box 9550, Berkeley, CA 94709. Telephone 800-227-0750. Or write: Information on Demand, Fairview Park, Elmsford, NY 10523. Telephone: 914-592-8134.

## DATASEARCH

Based in Sacramento, California, it offers a nationwide search of vehicle ownerships and driving records. The literature states: "Simply give us the full date of birth and full name, and we obtain the driving record." Other services are available, including occupational licenses, financial responsibility information, along with voter registration data. Searches of public record, nationwide, are also listed as a capability. DataSearch corporate headquarters are located at 3600 American River Drive, Suite 100, Sacramento, CA 95864, or call toll-free 800-452-3282.

## PUBLIC INFORMATION RETRIEVAL SERVICE

A new sourcebook first published in 1993, updated to 1995, is a national guide to firms who pull files and documents from U.S. and local courts and county agencies. Record *Providers* search by networking and databases. Record *Retrievers* go directly to the files. Public Record Research Library is published by BRB Publications, Inc. Its title: Local Court and County Record Retrievers. Probably available at your friendly neighborhood law library, or staff there will know where you can find it. A valuable volume, one well worth hunting for.

### INFOSEARCH, INC.

Infosearch offers "complete public information and document retrieval anywhere" and is capable of dealing with 60,000 separate jurisdictional locations in the United States, each with separate sets of rules for information release. A tall order. InfoSearch checklist of services includes, under "Miscellaneous," copies of birth, marriage, and death certificates, and probate and bankruptcy searches. All supplied, literature says, through "high-technology methods." You can write for more information to Infosearch, Inc., 1455 Response Road, Sacramento, CA 95814. A toll-free call: 800-222-2248.

### TRW CONSUMERS RELATIONS

One of the biggest and most used credit service agencies in the United States, this is an unlikely subdivision of TRW Space Hardware. It is unqualifiedly the most thoroughgoing of all credit facilities. Information is attainable from TRW Consumer Relations if you can supply the subject's full name, including any "Junior" or "Senior" or III or IV appellations; date of birth; spouse's name; Social Security number; present and previous addresses over the past five years—with zip codes. Send along a $10.00 money order with your request. P.O. Box 5450, Orange, CA 92667. Retrieval of credit records is available by letter only. All telephone numbers at TRW respond with a taped message citing the information required in your letter requesting credit card history. Which I've already told you about above, saving you a phone call.

## CONSUMER CREDIT REPORTING AGENCIES

Credit agencies as a rule maintain information on individuals for any preceding twelve months. Only credit union, business records, and mortgage specifics are not retrievable.

When you buy a suit or dress, some nice salesperson asks if you would like to open an account. The information on the credit application you're asked to complete may go to one of the many consumer credit reporting agencies, where a file is opened for you. This becomes a record of your full name, address, telephone number, employer, income, type of account, credit cards, credit references, account status, payment profile, and the account's number. In other words, the nice salesperson knows your personal financial history before he even hands your measurements to the alteration tailor.

And so it will go with every creditor you solicit who uses a consumer credit agency; you sign a credit release.

This information is intended to be confidential, and for the most part it is. It's available only to the creditor who subscribes to that agency's services, or to anyone for whom the subject has signed a release of credit information (a landlord or car dealer, for example), or perhaps to an agency *insider* who has the privilege of peeking at those credit files.

Now, let's assume you need a credit background on your subject, that suit/dress buyer, and don't know a credit agency file peeker. Where can you get a copy of his credit release and check him out? From him? Not unless he wants you to have it. From his landlord? H-mmm. Maybe your subject skipped the premises by moonlight and left behind a dog-stained carpet along with an Excedrin-size resentment. Or maybe your subject stuck the drugstore he patronized with $40 worth of snapshots, all of them taken at an Elks convention in Cleveland.

It's an all too painful fact for the typical American: he or she can't help leaving a paper trail of some kind. And if you are assiduous enough, and finding your subject is important enough to you, look for a creditor your subject has "burned." There you will find an ally in your search.

## CREDIT CARDS

Credit card companies have investigative and fraud divisions. Should you suspect that your subject is misusing your card number, you may get help by contacting the company. A few of the most prominent are listed here:

American Express                  World Finance Tower
                                      200 Vesey Street
                                      New York, NY 10285

Visa International                P.O. Box 8999
                                  San Francisco, CA 94128

MasterCard                        1 Custom House Square
                                  Wilmington, DE 19899

If your subject is in arrears or involved in fraud with *his* credit card, there's a good chance the company will be helpful in your search, depending on its intent and character.

## BUSINESS CREDIT REPORTING AGENCIES

Dun and Bradstreet, Standard & Poors, Moody's, and TRW rate and report credit on businesses and corporations. Access to their records is through subscription and membership, but is also available at some of the larger public libraries. These records contain background in depth about business owners and operators.

# MISSING PERSONS

## THE VANISHING ACT

An estimated 1.8 million people are reported missing in the United States each year. Most of them disappear intentionally, adults and children alike. They simply run away. Possibly 225,000, annually, never return home. Based on these estimated figures, there are millions of missing persons in the United States because they (1) clearly want to be wherever they are; (2) have no home, literally or figuratively; (3) are being held somehow against their will; or (4) are victims of homicide, suicide, accident, or unattended, maybe unrecorded, natural death.

The adults who flee are traditionally men. But the search organization Tracers of America reports it is now looking for as many runaway women as men. Men for the most part are escaping financial problems. The women—married, as a rule—seem to be running away from home in lieu of divorce, or from a situation a divorce wouldn't help or they can't afford.

The disappearing adult who doesn't change names is

relatively easy to find. We're trying to show you how to do that by following the paper trails every adult must leave in this day and age.

## THE NAME CHANGERS

Let's look at the name changers first.

The adults who *do* change names and go completely undercover with a new identity are relatively few, particularly among those with professions and trades who expect to earn a respectable living. Today's employer, government or private, seldom hires without researching the prospect's background. He requires references, licenses, training certificates, or other evidence of education or qualification for the job. And it isn't easy for the runaway to duplicate authentic information like this under an assumed name.

He or she could visit cemeteries and find a name on a tombstone with a date of birth corresponding roughly with his or her own. The runaway could obtain a birth certificate, apply for a driver's license and Social Security number, open a bank account—then try to find a job for which he or she can offer the employer only an oral résumé. If the runaway is a professional something-or-other and decides to go the identity-change route, the chances of winding up waiting on tables or pumping gas are pretty good. Probably not a very rewarding way to disappear.

On the other hand, the name changer can go the whole route, at considerable cost, and hire the services of one of the firms that supply documents on demand. Here an entirely new individual can be created on paper. New name, birth certificate, Social Security number, college degree, trade skill certificates. The runaway names the document and they'll be happy to *falsify* it. As a consequence, there is considerable risk of exposure somewhere down the line, not only causing the "new" person embarrassment but also assuring an excellent chance of going to jail for using fraudulent official documents. Such false document companies advertise in tabloids such as the *National Enquirer* and its clones.

My long experience, supported by statistics, says that choosing a small, remote town in which to disappear isn't an ideal way to go, either; the stranger is very likely to arouse friendly, but maybe dangerous, interest. Nor is it wise to elect to live in a place he or she has always talked about to friends and relatives and coworkers. A serious searcher—like a collector armed with an accurate Profile—would have no trouble finding his or her way to the subject's door.

Statistics and experience also suggest that the best place in which to try to get lost is a medium-to-large-size city—Tulsa, Oklahoma; Youngstown, Ohio; New York City; or Los Angeles, perhaps. The most suspicious towns in the world are the resort hangouts, populated with other name changers, law-enforcement officers, and bondsmen's bounty hunters.

In any case, changing identities is not something that is achieved casually. So you can see that the odds of finding your subject operating under his or her own name—or something close to it—are in your favor.

## THE MISSING-PERSONS REPORT

American law-enforcement agencies have an almost mandatory response to missing-persons reports, unless it involves a child: "He'll be back—give him some time." They're busy handling what they *know* has happened. But the ghastly Jeffrey Dahmer case proved the rule on vanished children.

Several missing-children reports in Milwaukee had gone uninvestigated when a woman called in to report a male adult abusing an adolescent male in the yard next door. Upon arrival at the scene, two inept city policemen admonished the adult, left the boy's cries behind, and drove away. Abandoned by the cops, the boy became one of the mutilated and cannibalized horrors found in Dahmer's apartment. Wisconsin has no death penalty. But Dahmer lost his life anyway. He himself was cut to death by another prisoner soon after he began serving a life sentence.

Indifferent missing-persons follow-up proved embarrassing to Chicago police. John Wayne Gacy killed and buried thirty-three young men and boys under and around his house over a period of several years. Although a convicted sodomist, Gacy went uninvestigated by police in spite of repeated requests to police to do so by one of the victim's parents.

Jan Charles was an attractive young woman who found herself rejected by a boyfriend in the small town of Brandon, Florida. She met a married man and became involved with him and with drugs. One night she was observed entering a car in a restaurant parking lot. She failed to return home.

When the police declined to take a missing-persons report—Jan was of age and mentally stable—her mother took up a search that was to last four years. She began by asking police to investigate the married man Jan had been seeing, and got no action. So Peggy Charles did her own sleuthing—in bars, cruising the highways, walking the streets of surrounding towns. Periodically, she asked the police for help, but none was forthcoming. The married man eventually went to prison on drug charges, and the police at last began an investigation that confirmed the man's involvement in Jan's death. In a plea-bargain that would allow him to serve a manslaughter charge concurrently with the drug sentence he was already serving, he led police to Jan's buried body.

These are just a few examples of official mishandling of missing-persons reports. It will come as no surprise that there are thousands more such cases, and with good reason: lack of time. Almost any law-enforcement agency today works overtime on caseloads that demand immediate, urgent attention in order to maintain a day-to-day commitment to keeping the peace. There is little time left over to deal with emotional, lengthy, and often inaccurate reports of "disappearances," most of which will resolve themselves within twenty-four hours.

Only a small percentage of those reported missing really are missing. Most turn up again within days, many of them within hours, especially children—whose reported disappearances, incidentally, get number-one priority from most police departments. But remember that .00434 percent of 250,000,000, the population of the United States, amounts to plenty of people who *stay* missing each year.

## THE NCIC AND MISSING PERSONS

Richard Ruffino is the former head of the New Jersey State Missing Persons Commission and a widely respected authority on the vanished. He contends that *all* missing persons should be recorded in the computers at the FBI's National Crime Information Center (NCIC) in Washington, D.C., where they do have the time and the personnel to handle such traffic. Ruffino says: "Let's give the benefit of the doubt to the relatives who want to report someone missing. It takes only minutes to enter a name, minutes to delete it if it turns up."

Entry of missing adults into the computer is currently limited to the physically and mentally handicapped, the elderly, or those disappearing under suspicious circumstances. The system currently stores about forty thousand disappearance names, almost all of them children. And today, thanks to the Missing Children Act of 1982, parents can go directly to the FBI with requests to enter the names of their vanished kids into the NCIC computer system if the police for some reason can't or won't do it for them.

The NCIC computers are plugged into every law-enforcement agency of any consequence in the country. NCIC has the only existing system for alerting those agencies, simultaneously, to crime activity in any of the fifty states. But until 1984 it had been used only sparingly for missing persons reports.

NCIC has served primarily in the storage and retrieval of data on individuals, weapons, vehicles, and property involved in felonies. NCIC's computer system has been responsible for the apprehension of suspects and the recovery of property in thousands of cases, nationwide.

Now it will play a bigger and better role in the locating of missing persons.

## MISSING CHILDREN

To date, 1996, the missing children picture remains confused. Hundreds of thousands of hours are spent on the problem each year by volunteers and professionals concerned with the phenomenon of vanished children. But for several reasons none of the reporting agencies can claim any degree of accuracy in the estimates made of the number of kids who disappear.

The numbers come from the FBI, the NCIC, the National

Center for Missing and Exploited Children (NCMEC)—all of them operating under the direction of the Justice Department. Some gratuitous estimates come from outside law enforcement: members of Congress, assorted bureaucrats, television preachers, columnists, concerned parents of victims, disappearance watchers, and psychics—all of these sources contribute in one way or another.

Kidnapping estimates range from 150,000 annually (the figure cited by a militant father of a murdered victim testifying before Congress) to the 69 reported officially in 1986 by the FBI.

Much of the numbers-conflict arises from the confusion over nomenclature for the disappearance of a child. No one knows exactly what to call it at first when a child turns up missing.

It could be kidnapping (a stranger's taking the child by force or stealth from its lawful or natural custodial parents or guardians); it could be parental kidnapping; it could be an abduction (that is, child-stealing by members of one parent's family); or it could be a simple disappearance, a case of the child's running or wandering away, never to be seen again.

Neither the family reporting a missing child nor the law-enforcement officer to whom the report is given can be expected to know for a matter of many hours, or even days, the *exact* circumstances surrounding a child's disappearance.

Therefore, the "kidnapping" numbers problem begins at the beginning, with the difficulty in identifying the event.

## KIDNAPPING AND PARENTAL KIDNAPPING DEFINED

Kidnapping is traditionally defined as an overt child-stealing act followed by a demand for ransom, a definition used by the FBI because it makes a clear-cut case; knowing this lessens the apparent absurdity in contrasts between the Bureau's reported kidnapping figure of 69 for 1986 and official estimates of tens of thousands of disappearances of children.

The fact is that often children are kidnapped for more sinister reasons than money.

Parental kidnapping is rarely—maybe never—money oriented, neither is love for the child necessarily a part of the parental child stealer's motivation; jealousy, resentment, and revenge against the custodial parent are the common reasons here. It can mean defeat for law enforcement when the stealing parent's intentions carry him or her and the child far out of local jurisdiction.

In any event, there will be no way possible to accurately estimate the numbers of children who vanish permanently to unknown fates until a means is discovered to categorize and monitor the individual cases from the beginning.

The FBI has stated recently that most of the estimated 30,000 children reported missing every year are runaways who turn up the next day. Child-stealing by noncustodial parents, says the FBI, accounts for most of the other children reported among the longtime missing. The Bureau acknowledges that many disappearances—for whatever reason—go unreported.

## CONFLICTING KIDNAPPING ESTIMATES

Ellen Goodman, a syndicated newspaper columnist, states that Child Find, a private New York City agency, estimates that there are 600 kidnappings in the city each year. She goes on to say that some sources claim that "there are 1.5 million children missing in this country, and that two-thirds, three-quarters, or 90 percent (according to three different authorities) are runaways."

Further data offered by the National Center for Missing and Exploited Children indicate that the missing children figures are not being exaggerated. In Chicago alone, 15,604 persons under age seventeen were listed as missing in 1990. Only 8,000 of these were classified as runaways. In March 1993 Florida's Department of Law Enforcement reported 4,007 cases of children missing in that state as of that date. The startling national figure of 1 to 1.5 million missing children may not be all that misleading.

Regardless of the conflicting numbers, it appears possible that every year more children, for whatever reason, disappear in the United States than we lose to car accidents, heart problems, and cancer.

## TIPS FOR CUSTODIAL PARENTS

The following are suggestions from a handbook published by the National Center for Missing and Exploited Children.

- If your child becomes a victim of parental kidnapping, go to your local police department and file a missing persons report under the Missing Children Act (passed in 1982 to assure that complete descriptions of missing children are entered into the National Crime Information Center's computer system, even if the abductor has not been charged with a crime). You don't need a custody decree to file this report.
- Contact parental kidnapping support groups to help you through the process of finding your child. (The location of the support group nearest you can be had by calling NCMEC toll-free at 800-843-5678.)
- Obtain legal custody of your child if you have not already done so. You can petition the court for custody even after your child has been abducted. Consider hiring a lawyer to help.
- Consider asking the police or prosecutor to file criminal charges against the abductor if you intend to press those charges after your child has been returned. It is almost always necessary to have a custody order to press criminal charges.
- If the prosecutor charges the abductor with a felony, make sure the state felony warrant is entered promptly into the National Crime Information Center computers. If the abductor has fled the state to avoid felony prosecution, ask the prosecutor to apply to the local district attorney for an Unlawful Flight to Avoid Prosecution (UFAP) warrant. If such a federal warrant is issued, the FBI can then assist in the search for the abductor.
- Search for the child on your own as well as working with law enforcement. When you locate your child, immediately send a certified copy of your custody decree to the family court in the place your child is located. Then ask for law enforcement to help recover the child.
- If the police won't help without a local court order, petition

the local family court to enforce your custody decree. Ask your lawyer for help in this matter. Go back to your family court after your child is recovered to ask a limit to the abductor's visitation rights. Also ask the judge to add provisions to your custody decree to prevent a repeat abduction.
- Seek psychological help for your child if he or she is having a difficult time adjusting after the abduction.

## THE NATIONAL CENTER FOR MISSING AND EXPLOITED CHILDREN

The National Center for Missing and Exploited Children is a national clearinghouse that collects, compiles, exchanges, and disseminates information on missing children.

The Center is a private, nonprofit organization that is funded in part by a grant from the Office of Juvenile Justice and Delinquency Prevention, Department of Justice. (The Center is not directed by the Department of Justice, however.) Its "hotline" now operates on a twenty-four (24) hour basis pursuant to the 1988 amendments to the Missing Children Assistance Act.

Anyone seeking a missing child or wishing to contribute information by mail should write to: National Center for Missing and Exploited Children, 2101 Wilson Boulevard, Suite 550, Arlington, VA 22201. Telephone 703-235-3900 or 800-843-5678.

## MISSING CHILDREN HOT LINES

**National Center for Missing and Exploited Children**  800-843-5678
**National Runaway Hot Line**  800-621-4000
**California**  California Youth Crisis Line  800-843-5200
**Florida**  Missing Children's Help Center  800-USA-KIDS
**Illinois**  Task Force on Parental Abduction  312-421-3551
**Pennsylvania**  Children's Rights of Pennsylvania  215-437-4000
**Rhode Island**  Society for Youth Victims  401-847-5083
**Texas**  Child Search  512-224-7939

Most of these organizations operate independently and are not necessarily in touch with each other. But a call to the National Center for Missing and Exploited Children at 800-843-5678, or to the National Hot Line, will get you the number of a group operating in your area or in the area in which you think the missing youngster is located.

## IF YOUR CHILD IS MISSING

Unless they're old enough to be taught how, kids can't identify themselves when they're lost.

Let's pray it never happens, but if your child one day turns up missing, notify the police or sheriff immediately. Explain that you will meet them where the child was last reported seen, then go there. Take along the latest photo of your child. Talk to *everybody* at or near the scene, and write down license numbers of cars parked nearby. Take down the names, addresses, and phone numbers of anyone who offers information. Knock on doors of homes and talk to merchants in the area. Stay calm and businesslike, so you can judge the value of what you're told. Try your best to find and talk to the last known person to see your child.

Provide officers with a recent photograph and a detailed description of your child and the clothes he or she was wearing when last seen. And finally, tell the officers—no matter how seemingly trivial it seems—what you have learned by talking to witnesses and your child's friends.

# SAFETY TIPS

## SAFETY TIPS FOR CHILDREN

All children can be taught how to avoid trouble, how to spot trouble, and what to do if trouble happens to them. Not enough kids know these things.

Safety and crime prevention should be a family effort. It's your job to teach kids how to be safe.

Teachers, law-enforcement officers, and others who have some responsibility for the well-being of young children should also share the parents' concerns.

(Many of the suggestions that follow were taken from the book *Take a Bite Out of Crime—How to Protect Your Children*, published by the U.S. Government Printing Office, Washington, DC 20402.)

### TEACHING CHILDREN SAFETY RULES

Talking to children about certain dangers to their personal safety makes many adults uneasy. It's difficult and painful to acknowledge that there are people around who want to harm children. And some safety rules you'll teach your children may seem to contradict other values, like being polite and obeying adults.

Children have a right to know about potential threats to their safety and to be told what to do if they find themselves in trouble. Some basic knowledge will increase their confidence and self-reliance. And you'll feel better knowing that they know how to cope when you're not around. And more and more kids today spend time at home and at play alone while their parents are at work.

*Rule Number 1 for any adult caring for very young children: never leave them alone at home, in a car, or in any public area. Not even for a minute!*

It is very important, when teaching children personal safety and caution, that they don't become overly frightened. So never present too much information at once; young children can't absorb a lot at one time. Children should know, of course, that the reason for learning and following safety rules is because they're loved. Rules should not be seen as arbitrary, and adults should take care to be consistent with the rules they teach and practice what they preach.

### TEACHING CHILDREN HOW TO SPOT TROUBLE: BE ALERT FOR STRANGERS

Friendly strangers can be dangerous strangers. Strangers who offer treats to children may be offering threats instead.

Teach young children what a stranger is: *anyone they do not know well*.

Even if children know how to avoid trouble, sometimes trouble—in the form of dangerous strangers—may find them. Children will encounter strangers when you're not around. Most strangers are well meaning and not to be feared. But because children are trusting and vulnerable, they can fall for offers by adults who seem kind but are not. So make sure your children learn and follow these basic rules when you're not around.

■ Never accept rides, candy, gifts, money, or medicine from a stranger.

- Never get close to a car if a stranger calls out to you for directions or anything else. It is easy for a stranger to pull you into a car.
- Never give your name or address to a stranger.
- Never open the door to anyone you don't know.
- Never tell callers that you're home alone. Say your mom or dad can't come to the phone and will call back.
- Never volunteer family vacation plans or other information about your home.
- Always avoid strangers who are hanging around rest rooms or the playground and want to play with you or your friends.
- Never go into a house without your parents' permission.

*Lots of kids wear clothing or carry belongings that bear their names. Parents, remind your kids that people who call to them by name may still be strangers, and "stranger rules" apply.*

What about the persistent stranger? Here's what the experts say you should teach your kids:

- If a stranger in a car bothers you, turn and run in the opposite direction from the one the car is headed. It's not easy for a car to change direction suddenly.
- When frightened, run to the nearest person you can find—a police officer, a person working in a yard—or to a neighborhood house or store. While you should always stay away from strangers who approach you, it's OK for you to ask an adult you do not know for help.
- If a stranger tries to follow you on foot or tries to grab you, run away, scream, and make lots of noise. The last thing a dangerous stranger wants is a lot of attention.

Kids spend a good part of their lives at school—in the classroom and on the playground. Strangers who want to hurt children know this, too. Find out what the school's policy is for allowing children to leave school with adults other than a parent or guardian. Also find out what school security measures exist to ensure students' safety. Get together with other parents if you find these measures lacking or weak, and work together with school officials and local law enforcement to beef up school security. You won't be sorry!

## "PLAY IT SAFE"—TEACHING CHILDREN HOW TO AVOID TROUBLE

It is normal and natural that children will spend time playing or traveling out of sight of trusted and caring adults. The best way to keep trouble away from kids is to teach them to avoid areas and situations where trouble might lurk. Here are some basic "play it safe" rules for children:

- Talk with your parents.
- Never play in deserted areas such as the woods, a parking lot, an alley, deserted buildings, or new construction.
- Always stick to the same safe route in traveling to and from school or a friend's house.
- Always try to play or walk with friends. It's safer—and more fun!
- Never play or loiter in such public areas as washrooms or elevators.
- Try to wait with a friend for public transportation. Try to sit or stand near the driver on the bus.
- Always keep doors and windows locked when home alone.
- Never display money in public. Carry money only if necessary and keep it in a pocket until needed.
- *Never hitchhike. Never!*
- Never walk or play alone outside at night.
- Always tell a family member or other adult in charge where you'll be at all times and what time you'll be home.
- Always stay near and keep your parents in sight in a public place. If you are separated from your parents in a store, ask a clerk for help.
- Never get into a car or go anywhere with anyone without your parents' permission.
- If anyone makes you feel uncomfortable or afraid, tell your parents.

*Halloween poses special problems and danger for children. Some treats are tricks. Teach children to throw away any candy or food they may collect that was not wrapped and sealed by the candy company. Notify police if there are any suspicious treats. And remember, it's always best to accompany young children on any door-to-door activity.*

### CHILD MOLESTATION AND ABUSE

Let's face it. Almost all the rules and tips in this book about increasing children's personal safety have to do with your two big fears for children: sexual molestation and physical harm.

The average age of the sexually abused child is ten years.

Teaching "stranger rules" is smart, but not enough. In the majority of cases, the child's sexual molester is known to the child and the child's family. That's why only a minority of abuse cases is reported to the police. The abuser is often a parent, relative, baby-sitter, or close family friend. Children may give in to adults' sexual advances because they fear losing their love or fear their punishment. Therefore, they are especially vulnerable to sexual abuse by someone they know, such as a person who cares for them regularly. Children are trusting and defenseless. Make sure you check carefully the references of baby-sitters, day-care centers, and recreation leaders.

A child may not recognize sexual abuse when it happens, or even know it's wrong, especially if the abuser is someone the child knows. *Children must learn what appropriate "touching" is. Discuss it with your child.* Many children instinctively know what "proper distance" should be kept between them and other persons. Sometimes a child may be uncertain about the intentions of another person. In this situation, children should know that it's OK to respond in a way that makes them feel safe and more comfortable. Children usually know that genuine and gentle affection is different from someone who tries to touch their genitals or fondle them in any way that makes them feel unsafe. They should pull away immediately if someone suggests such actions, even if they're offered a present as a bribe.

### How to Respond to Children's Stories

Children often make up stories, but they rarely lie about being victims of sexual assault. If a child tells you about being touched or assaulted, take it seriously. Your response is very important and will influence how the child will react and recover from the abuse.

Stay calm. In a reassuring tone, find out as much as you

can about the incident. Explain to your child that you are concerned about what happened. Don't be angry. Many children feel guilty, as if they had provoked the assault. Children need to be reassured that they are not to blame and that they are right to tell you what happened.

A child may need to be taken immediately to a doctor or an emergency room. Try to be prepared to explain *exactly* what took place.

Law enforcement, special hot lines, or a child welfare agency should be contacted right away.

Sometimes a child may be too frightened or confused to talk directly about the abuse. Be alert for any *change in behavior* that might hint that the child has suffered a disturbing experience.

- Is the child suddenly more withdrawn than usual, refusing to go to school or afraid to be alone?
- Is the child having trouble sleeping, waking up with nightmares, or wetting the bed?
- Is the child complaining of irritation of the genital areas?
- Is there blood or physical injury in these areas?
- Are there signs of increased anxiety or immature behavior?
- Does the child show a marked change in behavior toward a relative, neighbor, or baby-sitter?

Listen when your children tell you they do not want to be with someone; there may be a reason. Listen carefully to your children's fears.

### OTHER CHILD ABUSE

Other than sexual abuse, child abuse may include physical violence, emotional cruelty and deprivation, and physical neglect.

Child abusers are persons usually known to the child. This means most cases aren't reported to authorities, and children continue to suffer because abusers are repeat offenders.

Child abuse is dangerous and against the law. Many abused children will grow up and victimize their families, and others, later in life. It is your duty as a citizen to report suspected cases of child abuse by contacting a special hot line, the police, or child welfare agency immediately. The children need help and treatment as soon as possible.

### TEACHING CHILDREN HOW TO RESPOND

How children respond to trouble will depend upon their age and the particular circumstances they encounter.

While it is important for a child to know how to avoid and spot danger, it is also critical that a child knows how to respond quickly and wisely when confronted with trouble. Children should understand that there are many people they can depend on and should turn to when they feel unsafe.

Teach children that the police are friends whose job is to protect them. If a police officer can't be located easily, a child should also know to run to or seek out a trusted teacher, a neighbor, or a friend's parent when frightened or feeling endangered. *Children should know that they should report trouble right away.*

Teach children how to operate the telephone to call for emergency assistance.

- They should know how to dial 0, 911, or other emergency numbers used in your area.
- They should memorize their area code, phone number, and address—and maybe a friend's number and address as well.
- They should memorize your work number.

Keep a list of emergency phone numbers—such as the fire department's and a close relative's or friend's—posted near all the phones in your house.

Walk the neighborhood and the route your child travels to and from school with your child. Point out places to go when in trouble—homes or certain stores—and unsafe areas to avoid.

Children can be prepared to respond to trouble through role-playing. Make up situations and rehearse responses to increase the child's ability to act rationally and calmly.

No one likes to think about all the possible threats to a

child's personal safety and well-being. But a safe child is one who knows what to do when trouble happens.

*A child's best response to trouble: using common sense. Like knowing when to stick up for his or her rights, and when not to. Small children should not fight back when outnumbered by bigger youths who want to take their bike, radio, or other possessions. In this situation, a child should give in and then run to an adult, or older brother or sister, and report the incident right away.*

### SAFE COMMUNITIES: SAFE CHILDREN

To increase the safety of your children, increase the safety of your neighborhood. Keeping your neighborhood safe is the responsibility of concerned adults working together to prevent crime.

Start with your local law-enforcement agencies. They can tell you what special crime problems your community may have and what you can do about them. They can also tell you about other crime prevention or child safety programs already operating that you can join.

### Safe Homes

The Helping Hand, Block Parent, or McGruff Safe Home program is a good idea. If you don't have one already, start one. Here's what it's about.

Neighborhood parents, grandparents, or other adults are recruited to volunteer their homes to serve as temporary shelters for frightened or lost children. Volunteers can be trained to aid and comfort children. Special signs posted in the windows of their homes tell children this is a place to go when they are confronted with a serious problem. Local law enforcement can help in setting up a sound and effective program.

### After-School Programs

Millions of children are home alone after school while their parents work. Think about starting an after-school program—at school, in church, or in one of your community centers.

### CRIME PREVENTION CURRICULUM

More and more schools are adding personal safety instruction and crime prevention to their school agenda. Teaching children how to play it safe and what to do if they're threatened can and should be taught in all grades. The more often they hear that they can keep from being hurt, the less fearful they'll be. Children benefit from such instruction in the school and can learn a lot from exchanging ideas and experiences with their classmates. Kids can also learn ways to protect their property, skills to help them cope at home alone, and rules of good citizenship. Many of these classes are taught by the local police or sheriffs who bring along special puppets, coloring books, posters, and other teaching aids to reinforce their invaluable lessons. This way kids get to know personally the authorities who are there to protect them. Schools can help teachers get training in crime prevention techniques for children. Older kids are good teachers, too. Discuss these ideas with the PTA, school board, and local law-enforcement authorities. Crime prevention can also be a part of your child's organized extracurricular activities. Crime prevention clubs are a good way for kids to be a part of their community's crime prevention program.

### BLOCK PATROL

Organized patrols are another way of increasing child safety in the neighborhood. Parents, grandparents, and other concerned adults volunteer to be observers and reporters as they patrol the neighborhood during the hours children travel to and from school. Patrollers keep an eye out for trouble, record descriptions of strangers and their cars, observe potential traffic and other hazards, and report all suspicious activity to the police or sheriff and their neighbors. Being a block patroller doesn't require special skills or a lot of time—especially if you recruit a lot of volunteers to alternate duty. Patrollers prevent many tragedies from occurring. They increase the security of their neighborhoods and their children, and display their investment in both. Law-enforcement files are full of unsolved cases of missing children where there are no clues about the child's disappearance. Observant block patrollers could provide many such vital clues. Contact your local law-enforcement agencies to see if they have this program in your area.

### PARENT ALERTS

Sometimes a child who hasn't arrived at school isn't home sick or playing hooky. If something happens to a child on the way to school, parents often don't know until later in the day. Precious hours are lost to law-enforcement investigators when crimes are reported to them several hours after they occur. Many schools have organized volunteer parents and senior citizens who call the parents of absent students to check that the children are somewhere safe.

Lots of concerned parents are getting their children fingerprinted. Remember, fingerprints must always be retained by the parents or guardians, never by third parties. Fingerprinting will not prevent the tragedy of child abduction or disappearance, but it may help law-enforcement authorities identify children unable to identify themselves. Officials in North Carolina are putting microdots into the molars of some children. Tufts University has developed a technique for teeth prints. Some parents have their kids wear identification tags.

*A safe home and a safe neighborhood increase children's—and everyone's—safety. Kids imitate the actions of the adults around them. So get going on crime prevention—make your home secure, mark your valuables with an ID number, get the schools involved in crime prevention, and work together with others in programs like Neighborhood Watch. And always report crime or suspicious activity to law enforcement right away.*

## FOR MORE INFORMATION

Your local law-enforcement agency is a good place to start for more information. Law-enforcement agencies can tell you what programs are already in place in your area and help you get started in crime prevention. They can tell you which local civic organizations, scouting groups, human service agencies, and schools are involved in crime prevention and child protection.

Find out if your state has a crime prevention organization. If so, write and ask for information and assistance on crime prevention activities.

The organizations listed in Appendix 2, "Missing and Abused Children Organizations," can provide you with specific information on child protection, child abuse, and missing children.

### CHILDREN OVERSEAS

The U.S. Department of State can conduct a "welfare and whereabouts" search to locate a child and determine the physical condition of the child if he or she has been taken from the country. Such requests should be made to: Office of Citizen Consular Services, Room 4811, Department of State, Washington, DC 20520; or by telephone at 202–647–3666.

## SAFETY TIPS FOR ADULTS

### In Your Home

**Do:** If you lose your keys, wait until the last minute—within 48 hours at the most—to change your standard locks with dead bolts. Install a peephole in the front door if it's solid and hasn't got one. If there is a garage entrance to the house, be sure it and the outside garage door are secure.

Women—Use your last name and initial in the phone directory and on the mailbox or apartment door tags to avoid disclosing your gender.

List and file serial numbers of anything mechanical or electronic that can be carried away. Be sure to identify service people you've summoned by phoning the company they say they represent, when they arrive.

**Don't:** Never open the door to strangers. Leave your valuables within sight of strangers. Volunteer information about yourself, or give your whereabouts schedule to service people: "Don't come this afternoon. There won't be anybody here."

## Overnight or Holiday Absence

**Do:** Secure every door and window, upstairs and down. Leave a contact number with a friendly neighbor. Stop *all* deliveries, including newspapers and mail. Ask a neighbor to keep trash off porch and drive. Leave a light on upstairs and down, and a small radio tuned to a talk station.

Have a trusted neighbor change drape positions daily, and a kid to trim the lawn if you're on an extended absence.

**Don't:** Tag your keys with your name or address. "Hide" your keys under doormats, or on door frame. Leave notes for thieves to read and act on: "Bill, we've gone to Mom's for the weekend. Call us there." Leave your outside lights on. In daylight they send a signal to felons that there's no one home. Leave ladders lying around outside.

## In Your Car

**Do:** Use well-lighted, well-traveled streets at night. Travel with all doors locked, and—in winter—with the windows up, always with your seat belt buckled. Keep home and car keys separated so only ignition key remains with the car at attended parking lots.

Keep your driver's license number and VIN (vehicle identification number) handy.

**Don't:** Leave your valuables exposed in the car. Use your trunk. Leave credit cards, driver's license, or any important papers in the glove compartment.

## Breakdowns

**Do:** Day or night tie a handkerchief on the antenna and sit inside with your doors locked and windows closed and await help.

If you have a convertible, raise the top and stay locked in.

**Don't:** At any time accept a lift from a stranger.

## Parking

**Do:**  At night always park in well-lit areas.

Always lock your car and take your keys with you.

In parking facilities, leave only your ignition key with the attendant. Take your trunk and house keys with you.

**Don't:**  Park leaving your keys hidden in your car.

If you are being followed do not go home. Drive directly to a police station—or, if that is not possible, an open business—and report the incident along with the auto license number to the police.

OUTDOORS

## Personal Security

**Do:**  Use well-traveled, well-lit streets.

Be alert—watch for shadows.

If attacked, use any available object as a weapon, such as an umbrella, shoes, or even keys.

If in a crowded area—scream.

If you are being followed, cross the street. If the person follows, run to the nearest residence and telephone the police.

Be alert—pickpockets work in pairs; one jostles the victim while the other extracts the wallet.

Carry small amounts of cash.

Women: Always make sure you are carrying your handbag with the clasp facing toward you.

**Don't:**  Walk alone after dark.

Shortcut through dark areas.

Stand near doorways, shrubbery, or clumps of trees.

Accept rides from strangers.

Open your wallet so a stranger can see your identification or how much money you have.

Women: Leave your handbag in an unattended location or carry it open.

If you are a victim of crime, always press charges. Criminals need to be punished. Keep the telephone numbers of the po-

lice and fire departments and your apartment's security officer in a convenient location so you always have them on hand.

The "Safety Tips for Adults" on the preceding pages are
from one of the pertinent pamphlets available by writing:
Superintendent of Documents, U.S. Government Printing Office, Washington, DC 20402. Write and ask for copies of current pamphlets regarding personal and home security.

# THE ADOPTEE/
# BIRTH PARENT SEARCH

Without question, looking for one's natural parents or a birth child is the most poignant of all searches, and procedures required here are more complex than those used in searches for other subjects.

Some of the problems connected with the adoptee/birth parent search are caused by official attitudes left over from the last century. You will still find records clerks who stiffen with resistance at the very mention of the word *adoption*. Antiquated laws exist to this minute that deny an adult adoptee the privilege of searching his or her birth records; documents that are available on request to any American who can claim blood-line parents.

It isn't possible here to analyze the particulars of each state's adoption legislation. They vary widely; but Appendix 9 will help you in getting to know to whom you can write for official information, and realistically expect to get an informed response.

Florence Fisher, founder in 1971 of the Adoptees' Liberty Movement Association (ALMA)—now with more than 100 U.S. chapters—has stated in her speeches with some bitterness:

*In the United States the adopted adult at any age is denied the right to see the records of his birth and adoption. The adoptee must show "good cause" why he or she should be shown those records. Thus far, the courts have never deemed a man's or woman's desire to learn about the truth of their origin good enough cause! This sealing of adoption records, often at birth, is an affront to human dignity, and perpetuates rather than eliminates the stigma of illegitimacy. We (adoptees) are granted conditional equality, i.e., the name of our adoptive parents and the right to inherit from them, while our natural heritage is buried, obliterating our rights as human beings for all time. These sealed records not only deny us our civil rights as citizens of the United States, but our human right to know our roots. Where else is such secrecy condoned? Apart from slavery there is no other instance in our law in which a contract made among adults can bind the child once he reaches his majority.*

## THE NEED TO KNOW

The justice of a psychological "need to know" for adult adoptees and for birth parents of relinquished children is seeping into American courts. So is the urgency of adoptees' need to learn their *biological* heritage and the genetic forces that can mean the difference between health and illness or even life and death.

This new understanding seems to be filtering down to the heretofore superstrict recordkeepers who watch over "orphan courts."

You can take advantage of this refreshing attitude. I don't have a long history of success in this area to point at, but my research shows that you can get results from these court sources by mail and telephone. In the pages ahead, I provide you with the names, addresses, and in some cases, phone numbers of public and private agencies in fifty states equipped to help you acquire answers to "Who am I?" or "Who and where is my child now?"

Prepare your communications carefully, follow my instructions, persevere. And you will learn what you must know.

## BEGINNING NOTES

### STEP 1

Think. Your search for a birth parent or relinquished child could make changes in your life that might have a possible shattering effect on others. If you are successful, are there those who will be hurt? Whose love and trust might you be about to abuse? Is there someone who will be embarrassed or infuriated? Of course, you have considered these consequences many times. You wouldn't be a reader of this book if you hadn't arrived at some kind of resolution to accept whatever happens. I just want you to take one more *serious* look before you begin.

### STEP 2

Read this book from cover to cover, take lots of notes, and mark pages of special interest to you as you go. Collect the same information on your subject(s), from every source you can find, as I recommend in Chapter 1. Make a Profile of the principal subject(s), including those statistics upon which good investigations are built. Dates! Places! Names! Reread the beginning of Chapter 4! Finally, familiarize yourself with the kind of information available at *all* courthouses.

### STEP 3

Unless your search must be kept secret, get help! You need all the backup you can muster from friends, relatives, anyone you can interest in what you are doing. Choose a genuine cheerleader as your principal supporter. When slow or difficult going has you saying "What's the use!" you need someone who will say "Nonsense! We have just begun to fight!" Remember, you are scrapping for your own or your child's identity.

### STEP 4

I've asked you to create on paper the best image you can assemble of your subject(s)—a Profile.

Now I'm asking you to become a "scenario" writer. This

shouldn't be too hard. If you are adopted or have relinquished a child for adoption, you've had lots of practice filling in the blanks with your imagination.

In your scenario, use a "once-upon-a-time" style. Begin with "I was born" or "I gave birth," and go from there. Include every memory, every scrap of growing-up hearsay, probability, and conjecture that comes to mind. But if you can, make the *heart* of your story factual. Try to use real-name characters, and places and circumstances you are sure of, to help your own credibility: "I remember living in Akron. Then I was in a car, with my adoptive parents in the front seat, following a moving van to Detroit. I had my sixth birthday on the trip and a waitress in a diner brought me a cupcake with one candle on it. Later my little adoptive sister Amy threw up on our lunch basket and got a spanking. The back doors of the truck we followed had a big world painted on them." Believe it or not, practicing recall of details like these leads to other, more important scraps of memory that together could add up to a critical page in your book of life.

Maybe you can still see yourself in the recovery room after the baby boy you called your little David was born; maybe you can still feel the anguish you felt signing the adoption papers for the kindly judge with a white mustache; and maybe you remember wondering what life would be like for your little boy with the strangers you would never see. Put it all down—your emotions, places, names, impressions. This is all fabric onto which you can begin to sew the black-and-white facts you will find as your search goes on.

### STEP 5

Follow steps 2, 3, and 4 from "Beginning Notes" in Chapter 1, and plan to use the forwarding postcard device when the situation calls for it. For other correspondence when you are asking for information to be mailed to you, enclose a self-addressed, stamped envelope.

### STEP 6

Don't become discouraged! Expect frustration and delays. Remember that your communication could be one of dozens that cross the desk of an executive bureaucrat on any given day, and that it may be "handed off" to a subordinate for action. This consumes time, so delays are to be expected. But the wait will be worth it!

### STEP 7

Rally your support—and get to work!

### SUPPORT—MORAL AND PRACTICAL

In the human experience, the adoptee/birth parent search is one of the activities most deserving of moral support. The searcher need only look around to find everything he or she needs. But even the brightest and most loyal pal can't be expected to provide experienced guidance.

Curiously, this help could come from strangers, people you may never meet in person but who can help you because they have "been there." I'm talking about those who already have faced the clerks, the courts, the cop-outs by officials who won't go out of their way to help.

Dozens of organized and semiorganized volunteer support groups have been developed to guide those looking for birth parents or relinquished children. All of them are helpful. Some have reached the status of professionals in matters of adoption law, procedures, and uncovering information sources. And because all these factors vary so widely from state to state, these people could possibly save you weeks— even months—of research. Nearly all of these groups ask a modest membership or service fee. Seek out the support groups in your community or region. Or contact one of the following organizations. Let them know your circumstances, your plans, your problems, and stay in touch with them. They will help you avoid confusion, and they will give you an orderly step-by-step procedure to follow right from the beginning.

I suggest you select and join one of these groups before proceeding with your search. The cost is not exorbitant, and the help you get will be invaluable.

The following are some of the most important support

groups available to you. There are many more, and you can learn about those in your region by contacting a local adoption center.

NATIONAL ADOPTION SUPPORT GROUPS

Adoptive Families of America
3333 Highway 100 North
Minneapolis, MN 55422
612-535-4829

Adoptive-parent organization knowledgeable in intercounty adoption. Free literature on available agencies.

Children Awaiting Parents (CAP)
700 Exchange Street
Rochester, NY 14608
716-232-5110

The CAP book presents photographs and brief descriptions of backgrounds of children across the nation waiting to be adopted.

Child Welfare League of America
440 First Street NW Ste. 301
Washington, DC 20001
202-638-2952

Child-welfare group that researches and publishes books and literature on child welfare. Information provided on member agencies.

Committee for Single Adoptive Parents
P.O. Box 15084
Chevy Chase, MD 20815

Write for helpful information on single people interested in adopting opportunities. Publishes *The Handbook for Single Adoptive Persons*.

International Concerns Commitee for Children
911 Cypress Drive
Boulder, CO 80303
303-494-8333

Information on adoptable American and foreign children; an offshore orphanage sponsorship program; annual report on foreign adoption.

National Adoption Center
1218 Chesnut Street
Philadelphia, PA 19107
215-925-0200/800-TO-ADOPT

Registers information on both children requiring special care and adults interested in adopting them. Helps with interagency placements.

National Adoption Information Clearinghouse
1400 Eye Street NW, Ste. 1275
Washington, DC 20005
202-842-1919

A clearinghouse that offers easy accessibility to information on infant and intercountry adoption of children with special needs. Limited to referrals. Does not place or counsel. Available: *The National Clearinghouse Directory* listing adoption agencies, parent-support groups, adoption exchanges, and legal resources.

National Committee for Adoption
1930 17th Street NW
Washington, DC 20009
202-328-1200

Private adoption agencies organization monitoring legislation, counsels on pregnancy, infertility, and adoption. Available: *The Adoption Fact Book*.

National Resource Center for Special Needs Adoption
P.O. Box 337
Chelsea, MI 48118
313-475-8693

Publications, training, technical assistance on adoption
of special-needs children; services both individuals and
organizations.

North American Council on Adoptable Children
1821 University Avenue Ste. N–498
St. Paul, MN 55404
612-644-3036

Annual conference and clearinghouse for adoptive-parent
support groups. Emphasis on children awaiting adoption.

### ADOPTEES IN SEARCH (AIS)

This is a full-fledged organization near the nation's capital
that offers workshops, regular meetings, and dissemination of
search information. Its brochure states that AIS was formed to
aid adoptees, adoptive parents, birth parents, and siblings in
support of searches and in providing information concerning
adoption procedures. Membership is nationwide and is open
to those eighteen years or older. Younger members are allowed, with their adoptive parents' written consent. If you
live outside the Washington, D.C., area, you can apply for listing in the AIS birth registry.

To get a membership application and more information,
write Adoptees-in-Search, Inc., P.O. Box 41016, Bethesda, MD
20824, or call 301-656-8555.

### ADOPTEES LIBERTY MOVEMENT ASSOCIATION (ALMA)

ALMA is probably the strongest and most active of all
adoptee groups. Becoming a member places you on the
ALMA International Reunion Registry and newsletter-mail-
ing lists.

Adoptees over eighteen are welcome, along with birth

parents who have relinquished children for adoption, with the understanding that those children cannot be searched for until they are eighteen. Also encouraged to become members are foster children over eighteen years of age and adults separated from siblings by divorce, abandonment, or other causes.

The ALMA staff is volunteer, and all have completed their own searches. Although you do your own searching, ALMA is structured so that the staff can guide and assist you through proven search methods. Its literature offers "personal attention and understanding."

Following is a list of benefits included with ALMA membership in forty chapters nationwide:

**ALMA registry:** a computerized, multilevel, cross-reference system at national headquarters. It has more than a half-million entries.

**Search assistants:** Members who have completed searches offer time and know-how along with moral support at all ALMA chapters.

**Search workshops:** In Southern California these are held every month. Locations and times are announced in the newsletter.

**RAP groups:** Smaller groups can discuss emotional aspects of searches and get support in a casual setting.

**Searchers guidebook:** This is the official ALMA publication that takes members on a step-by-step trip through completion of a search.

You can apply for membership at any ALMA chapter or by writing National Headquarters, P.O. Box 727, Radio City Station, New York, NY 10101-0727.

### ADOPTION SEARCH INSTITUTE (ASI)

By and large, this is a *teaching* organization. Founded by Pat Sanders in 1980, it's an organization that was originally named Independent Search Consultants. Sanders quickly learned that search techniques could not be provided by her inexperienced volunteers. As time passed and enrollment grew, she developed classes in how to go about the search in an organized and effective manner. These classes provide the searcher with concise methods for retrieving information through records searches, and where to look for those records. Search classes are conducted on a regular basis for minimal fees.

Most important to the mail correspondent, ASI is ideal for referral during the many steps of any search. All of the volunteers you might contact by mail are trained at offering counsel.

Write: Adoption Search Institute, P.O. Box 11749, Costa Mesa, CA 92627.

### CONCERNED UNITED BIRTHPARENTS (CUB)

CUB dates from 1976. It is not essentially a search group. Its main function is to provide "inside" information, gained through the experience of its volunteer members.

CUB describes itself in a pamphlet titled "The Birthparent's Perspective" as "an autonomous national nonprofit support and advocacy group for parents who have surrendered children for adoption."

It also states that it is "the only nationally organized group available for those who research the dynamics of teen pregnancies and untimely pregnancy in adults."

Those welcome as members: adoptive parents, birth parents, and those interested. There is a newsletter, and there are monthly meetings. Write CUB, 2000 Walker Street, Des Moines, IA 50319. Or phone: 800-822-2777

### KANSAS CITY ADULT ADOPTEE ORGANIZATION (KCAAO)

So far, KCAAO can claim assistance in more than 600 reunions. Training and assistance are confined to those searching for an adult. The organization maintains a registry with more than 1,000 members, is listed and works with the International Soundex Reunion Registry and with ALMA's international Reunion Registry Data Bank.

Monthly meetings are held, and when appropriate, a newsletter is published for members' benefit. You can register at any age, but membership applications are not processed until you reach majority.

KCAAO was incorporated by the state of Missouri in 1979, and its statement of intent reads, in part: "Conducting training programs and seminars to teach (a) research techniques required to discover family history, genealogy, and adoptive facts; (b) methods (to) record information found; (c) the techniques of handling sensitive information."

Write Kansas City Adult Adoptees Organization, P.O. Box 15222, Kansas City, MO 64106. Or telephone 816-356-5213.

### ORPHAN VOYAGE

This organization prefers to operate as individual workers rather than in a group, although there are Orphan Voyage groups here and there across the country. Jean Paton founded Voyage in 1953, which makes it the oldest of the support organizations in the country. Birth parents and adoptive parents with children under eighteen are eligible.

Write Orphan Voyage, National Headquarters, 2141 Road 2300, Cedaredge, CO 81413; telephone 970-856-3937.

### TRIADOPTION LIBRARY

More than 300 adoption groups contribute to the support of this library, whose adoption literature has been gathered by its founder, Mary Jo Riller. This includes articles and books on adoption and relinquishment, recent law changes, and a reservoir of genealogical research references. Also referred are record searchers, psychologists, and consultants. There is no charge for taking advantage of this wealth of material. Questions will be answered by telephone or mail. You can become listed as a benefactor with a gift of $1,000.

The library itself is located at the Westminster Community Service Center, 7571 Westminster Avenue, Westminster, CA 92683. For more information, write: Triadoption Library, P.O. Box 5218, Huntingdon Beach, CA 92646. Telephone: 714-892-4098

Aside: It's notable with all the foregoing, and what follows, that *volunteers* make up the body of "employees" of the various support groups. This is a clear statement that those involved have discovered a sort of "mission," with the volunteers' own personal experiences a powerful motivating force to help others.

It also speaks of the efficiency that has developed within the groups. In spite of being volunteers—and thus amateurs in the pursuit of involved kinds of information—they have found a way to develop expertise that works. Like all volunteers—at anything—they must be content with the special wages of self-satisfaction—the sense of helping others through troubling periods in their lives.

Right here, a line or two from me to congratulate these unsung contributors to our national well-being. "Volunteers" are an American phenomenon that has helped our nation survive since 1775.

### REUNION REGISTRIES

Reunion registries are, for the most part, offered by adoptee and birth parent-support groups. They allow the matching up of identifying vital statistics submitted by both the searcher and the searched-for. When a match occurs between received information and information already on file, each of the parties is notified. Many people who have relinquished children, have been adopted, or who have misplaced siblings can thank the registries for their reunion. And *you* shouldn't overlook the good possibilities here.

The required information is supplied to the registry on an application form for which you must send. You may have to pay a modest fee, or the registration may be included in the support group's membership fee.

Most of the existing registries were established in the 1970s or 1980s. Word of their effectiveness has grown along with their rosters of search names. Most registries are private, but a few operate at the state level as agencies. So far, only California, Maine, Michigan, Minnesota, Nevada, Oregon, South Carolina, and Texas have set up people-matching facilities. Other states have begun to follow their example.

A regional listing of private registries follows.

## ADOPTEES' LIBERTY MOVEMENT ASSOCIATION (ALMA)

ALMA's 191,000 registrations of searchers and searched-fors, and its 48 chapters, make it the largest registry in the United States. Adoptees' Liberty Movement Association membership includes registration in the International Reunion Registry Data Bank. Make a note that although anyone can join, no search will be conducted for anyone under the age of eighteen. You can apply for membership at any ALMA chapter, or by writing the National Headquarters at P.O. Box 727 Radio City Station, New York, NY 10101-0727. Telephone: 212-581-1568.

## CONCERNED UNITED BIRTHPARENTS (CUB) REUNION REGISTRY

CUB has about 18,000 names in that category registered for hopeful reunions. It is the largest of the birth parent support groups. There is a registration fee in addition to the annual membership fee. CUB is not literally a search group, but works closely with those groups that are. CUB also expresses interest in the prevention of teenage pregnancies and has published informative literature on the subject.

To join, write: CUB, National Headquarters, P.O. Box 573, Milford, MA 01757.

## INTERNATIONAL SOUNDEX REUNION REGISTRY (ISRR)

ISRR uses a computer-ready system that translates all the information you submit on your application into one code line. The code line contains seven points of match, which permits match-checking of up to 1,000 registrations in five minutes or less. ISRR has more than 12,000 registrations and can involve more than 175 support groups in search efforts. It also offers two special services: one for those with medical problems, and one for those individuals who resulted from, or were donors of, artificial insemination. The former are asked to mark their application with bold lettering, MEDICAL ALERT, and to accompany it with a letter describing their infirmity. The latter should mark their applications with AID (Artificial Insemination Donor) or AIP (Artificial Insemination Product). These applications receive special processing at ISRR.

For registry in ISRR, write: P.O. Box 2312, Carson City,

NV 89701. Include a self-addressed, stamped envelope with your letter of inquiry. Membership is free. Telephone 303-494-8333.

### ORPHAN VOYAGE REUNION REGISTRY

One of the oldest registries, Orphan Voyage Reunion Registry has about 3,500 on its rolls. Adoptees under the age of eighteen are permitted to register if they have the consent of the adoptive parents. You must join Orphan Voyage to qualify for registration, and membership entitles you to not only registration but advisory help as well. Anyone of any age can join. You can get an application form and further information by writing: Orphan Voyage, 2141 Road 2300, Cedaredge, CO 81413; or telephone: 303-856-3937.

*The following is a listing of state-operated registries.*

### MAINE REUNION REGISTRY

This registry specifies that *both* adoptees and birth parents must want to be reunited. It doesn't provide search assistance for other missing relatives, and confines itself to notification of both parties if a match occurs. Only those over age eighteen can apply for registration. Write: Office of Vital Statistics, Department of Human Services, Station 11, Augusta, ME 04333; or telephone: 207-289-3181. There is a fee.

### MICHIGAN REUNION REGISTRY

Created by Public Act 116, this registry became effective on September 10, 1980. The date is significant because information-release procedures differ, depending on whether the relinquishment took place before or after that date. The registry is more concerned with its files of consents and denials for information release than in actual searches. Write: Department of Social Services, Adoption Central Registry, P.O. Box 3007, 300 S. Capitol Avenue, Lansing, MI 48909.

## MINNESOTA REUNION REGISTRY

The Minnesota Reunion Registry takes a more open view of the adoptee/birth parent searcher's needs. This agency cooperates with a number of volunteer organizations and refers searchers to them for help. ALMA, CUB, Kammandale Library, Liberal Education for Adoptive Families—all remain in contact with the state unit. These organizations in turn take the state's efforts into account when holding regular joint meetings, issuing newsletters, and so on. If you are twenty-one years old or older, you can request information on your original birth certificate from the state registrar at the address below. The registrar's office will undertake a search for your birth parents, if that is your wish. If they are located, they are notified of your request—confidentially. They then will file their consent or refusal of your desire to contact them. If they consent, you get their identifying information. If they do not, your birth certificate remains sealed and out of reach except by court order (for which you can file).

For information and an application for registration, direct inquiries to: State Registrar, Vital Statistics, Department of Health, 717 Delaware SW, Minneapolis, MN 55440; or telephone: 612-296-5316.

## NEVADA REUNION REGISTRY

This registry requires a notarized letter of identification requesting registration and giving permission for the release of submitted information. Write: State Reunion Registry, Department of Human Resources, 251 Jeanell Drive, Capitol Mall Complex, Carson City, NV 89710.

## NEW JERSEY REUNION REGISTRY

This is actually a registry of authorizations for release of adoption records. A notarized letter will get you listed; include the names of the people involved with your adoption on both sides. After the other parties have filed a similar letter, the information will be released. Write: Bureau of Resource Development, Box 510, Trenton, NJ 08625.

## TRACING BIRTH PHYSICIANS

The success of your search could well turn on the availability of your birth records. And bureaucracy everywhere being what it is, these can be sealed to you and open to somebody else. Like your birth doctor. He can review hospital and medical records at will, and he is not legally bound to keep them confidential.

Chances of finding these records naturally diminish with time and with the relative inefficiency of recordkeeping in bygone years. But it is helpful to know that *medical associations* run a tight ship in keeping track of their members. So if the identity of your birth doctor becomes critical, and he is still alive, these organizations can help find him, and he in turn can request and get the information you need. (See Appendix 3 for the addresses of state medical boards in the late 1990s.)

## PETITIONING THE COURT

It is important to bear in mind that if you are refused critical information regarding your adoption or child relinquishment, you have the right to petition the court to have those records made available to you.

If you can afford an attorney, it's best to engage one to file the petition for you. In many areas, paralegal help—the law's equivalent to a paramedic—is obtainable at low cost. Or your county bar association may offer volunteer lawyers' time free of charge. Your yellow pages will list this last service if it exists. And you can call the bar association itself to check on paralegal counsel. Remember, by law paralegals cannot offer concrete legal advice; they can only suggest how you may go about petitioning the court and they can prepare the petition for you.

In any event, you should approach your petition prepared to show "good cause," the nature of which will probably prove to be the strength or weakness of your case. Matters regarding inheritance rights make a "good cause." An even better one entails showing that your *biological* "need to know" is imperative to your physical well-being.

You will find the name of the appropriate court in your adoption state in Appendix 9 under "Court of Jurisdiction." A good first move here is to write the *clerk* of that particular court. Ask for proceeding instructions so that you can give your law counsel a starting point. It will save you both time in preparing your petition.

# GENEALOGICAL LIBRARY—THE CHURCH OF JESUS CHRIST OF LATTER-DAY SAINTS (MORMON)

### THE FAMILY HISTORY LIBRARY
(formerly the Genealogical Library)

The Family History Library of The Church of Jesus Christ of Latter-day Saints was founded in 1884. Its intent was and is to gather records for helping people trace their family lines. It is *not* designed for tracing the living and missing.

In 1938 the library began to record family history data on newly available microfilm. Today more than 155 microfilm cameras continue to record birth, marriage, death, military, land, probate, and other data in forty-five countries. The library has now accumulated the world's largest, most complete collection of genealogical information.

A library branch system was established in 1964. Now there are more than 11,000 of these vital statistics sources located throughout the United States, Canada, and thirty-eight foreign countries. Originally named the Genealogical Library, the main library in Salt Lake City has been known since August 1987 as the Family History Library. Branches throughout the world are now called Family History Centers.

The Family History Library in Salt Lake City and the centers everywhere are almost all open to the public. You may use these resources without charge, except for duplication and postage fees. The attending staffs are composed mainly of cooperative volunteers. And they can help you draw upon an incredible reservoir of information: 1.5 million rolls of microfilmed records are available at the Family History Library. Copies of these are lent upon request (mailing cost only) to Family History Centers for on-site viewing with projectors and microfiche.

There are more than 195,000 reference volumes on the shelves of the Family History Library, with a good sampling of these on microfilm or microfiche at the center nearest you.

Note that the volunteers are primarily guides to records, not genealogical experts. So you must rely on your own curiosity and ingenuity in following leads they may suggest in your search. When the volunteers can't answer your questions, they will help you fill out a reference form requesting the information from the Family History Library. Be brief with your questions, and give the specific localities, time periods, and records with which you need help.

A fifteen-minute slide presentation with a companion booklet will help introduce you to the library's system; it's available in English at the Family History Library and at most centers.

## Catalog

The Family History Library Catalog, for your use at any center, describes all of the library's records and lists them by number. Use it to find the book, film, or fiche numbers of the records you need.

## International Index

The International Genealogical Index lists more than 121 million births, christenings, marriages, and Latter-day Saints temple ordinance dates of deceased persons. The index is on microfiche at all centers, as well as at the Family History Library.

## Film Circulation

Most of the library's record resources are available for center use. There are modest postage and duplication fees for this service, which can be requested *only* through a Family History Center.

## Little-Used Files

Copies of microfilm records for the United States, Canada, the British Isles, Scandinavia, Germany, and the Netherlands are stored at the library. Microfilms of records for other countries are considered "little-used." They are not on file at the library.

If you have need of these and plan going to the Family History Library, you must write at least three weeks in advance. Give dates when you will be there, along with your name and phone number. Add either the film number or the specific localities, dates, and types of records involved. This advance notice is for your convenience. Requests made *after* you arrive in Salt Lake City will require a minimum of three days to fill.

## Copy Services

Photocopies of all records not copyrighted or otherwise restricted are available with self-help copiers at the library and many of the centers. At the library, the cost of each copy made from a book is 5 cents. Copies made from microfilm or microfiche are 20 cents each. These costs will vary at the centers. Where there is no copier, fill out a photoduplication order form and send it to the library. The form lists the costs for this service. There is a minimum charge of $2.00.

## Publications

For a guide that describes how to do genealogical searches in other specific areas of the world, visit your nearest Family History Center. The guide is free.

## Library Classes

Classes on genealogical research are held at the library on a regular schedule, a copy of which is yours upon request. Additional classes of ten or more people can be arranged and scheduled. Some centers also offer classes.

## The Family Registry

There are more than 185,000 individuals and family groups interested in sharing search information. They are indexed in the Family Registry, which can help you make contact with other searchers, perhaps those with information needs similar to your own. The index is on regularly updated microfiche at the library and most centers. You will be provided with a registration form for filing information, or when asking others in the Family Registry for help in researching a question.

## Ancestral File

The Ancestral File is a computerized database of information now being developed to help families share information. You can submit pedigree charts and your family records to: Family History Department, Ancestral File Operations Unit, 50 East North Temple Street, Salt Lake City, UT 84150; or telephone: 801-531-2584.

## Personal Ancestral File

The Personal Ancestral File is a genealogical software program for home computers. With it you can store and update information, and printouts will appear on family record forms, pedigree charts, and other forms. It is available only in MS-DOS and Apple MacIntosh formats.

For more information, contact the Ancestral File Operations Unit at the address above.

## The Professionals

To get the help of professional searchers, request a list of genealogists who provide their services for a fee. These individuals will have completed the Family History Library accreditation program, but are not affiliated with the library itself.

## Supplies

Pedigree charts, family record forms, and other supplies can be obtained at the library and at many centers. For additional supplies, check your local bookstores; visit or write: Salt Lake Distribution Center, 1999 West 1700 South, Salt Lake City, UT 84104; or telephone: 801-531-2504.

## Information Sharing

If you wish to share your genealogical findings with the Family History Library, they would be most welcome. Contact the Acquisition Unit of the library for more information on what can be accepted and how to prepare your material before submitting it.

## Prepare Yourself

Come prepared for your first Family History Library or Center visit. You will be more efficient and successful if you collect as much family information as you can beforehand, from records at home and from relatives. When you arrive, put what information you have on the pedigree chart and family group record forms you will be given. Then decide on the specifics with which you want to begin a search: a name, a place of birth, a critical date, and go on from there with your volunteer's help.

If you are planning to visit the library, before making the trip use the International Genealogical Index (IGI) and the Family History Library Catalog (FHLC) at your local Family History Center. It will save you time.

## The Rules

While the Family History Library and its Centers are private, the services are offered to the public with the understanding that users of FHL information must abide by guidelines and rules governing that use. These regulations will be explained by the library staff on the occasion of your first visit.

## Working Schedules

The working schedules of American and Canadian Family History Centers are available by telephone. Local centers are generally listed in your telephone directory white pages under The Church of Jesus Christ of Latter-day Saints Family History Center (or Branch Genealogical Library).

If you wish to contact the main library, write: The Family History Library, 35 Northwest Temple Street, Salt Lake City, UT 84150; or telephone: 801-531-2331.

Genealogical data is also available to the public through numerous public libraries in the United States. The New York Public Library has one of the most complete genealogical sections available in the United States. See Appendix 5 for a detailed list of branch genealogical libraries throughout the United States, Canada, and the rest of the world.

# WHERE TO WRITE FOR VITAL RECORDS

As part of its mission to provide access to data and information relating to the health of the nation, the National Center for Health Statistics produces a number of publications containing reference and statistical materials. The purpose of this publication is solely to provide information about individual vital records that are maintained only on file in state or local vital statistics offices.

An official certificate of every birth, death, marriage, and divorce should be on file in the locality where the event occurred. The federal government does not maintain files or indexes of these records. These records are filed permanently either in a state vital statistics office or in a city, county, or other local office.

To obtain a certified copy of any of the certificates, write or go to the vital statistics office in the state or area where the event occurred. Addresses are given for each event in the state or area concerned in Appendix 8.

To ensure that you receive an accurate record for your request and that your request be filled with all due speed, please follow the steps outlined below for the event in which you are interested.

1. Make sure you write the appropriate office.
2. For all certificates send a money order or certified check; cash lost in transit cannot be refunded.
3. Type or print all names and addresses in the request.
4. Give the following facts when writing for birth or death records:
    a. Full name of person whose record is being requested.
    b. Sex and race.
    c. Parents' names, including maiden name of mother.
    d. Date of birth or death—month, day, and year.
    e. Place of birth or death—city or town, county, state, and name of hospital, if any.
    f. Purpose for which copy is needed.
    g. Relationship to person whose record is being requested.
5. Give the following facts when writing for marriage records:
    a. Full names of bride and groom (including nicknames).
    b. Residence addresses at time of marriage.
    c. Ages at time of marriage (or dates of birth).
    d. Month, day, and year of marriage.
    e. Place of marriage—city or town, county, and state.
    f. Purpose for which copy is needed.
    g. Relationship to persons whose record is being requested.
6. Give the following facts when writing for divorce records:
    a. Full names of husband and wife (including nicknames).
    b. Present residence address.
    c. Former addresses.
    d. Ages at time of divorce (or dates of birth).
    e. Date of divorce or annulment.
    f. Place of divorce or annulment.
    g. Type of final decree.
    h. Purpose for which copy is needed.
    i. Relationship to persons whose record is being requested.

# IN CONCLUSION

My travels these past forty-odd years as a professional searcher have shown me the world several times over. And— it seems to me now—half the people in it. Repeatedly, I was taught that officials with whom you will be brought into contact by mail or telephone are *people*. You must understand that many of them would be happy to swap problems with you if they could. Their official attitude toward you is the professional performance for which they are paid. Just be aware that they can be moved by your sincerity to give you the information you need. Their in-baskets are full of needs like yours. But give them a warm, understanding-their-problems tone in your voice or letter, with honesty they can relate to, and you can move mountains—of paper.

Again, the three basic rules.

1. Persistence: don't be put off.
2. Consistency: state your exact needs.
3. Clarity: state your rights.

If you can fight off the inevitable frustration that comes from dealing with a seemingly disinterested official America, you will learn a priceless lesson. Nowhere else on earth does the phrase "freedom of information" mean what it does here. And to have to *scrap* for that information, occasionally, means that information about *you* is being protected from professional public record ransackers.

This book may not contain all the answers each of you needs. Such a book would require the pedestal you've seen supporting your library's ten-pound master dictionary. But I can say I truthfully believe that the determined searcher will discover in *How to Locate Anyone Anywhere* all the strategic "clues" necessary to plan and carry out a successful person-to-person search program.

# Appendix
# 1

# Social Security Index of Valid Numbers

| | | | |
|---|---|---|---|
| 001–003 | New Hampshire | 387–399 | Wisconsin |
| 004–007 | Maine | 400–407 | Kentucky |
| 008–009 | Vermont | 408–415 | Tennessee |
| 010–034 | Massachusetts | 416–424 | Alabama |
| 035–039 | Rhode Island | 425–428 | Mississippi |
| 040–049 | Connecticut | 429–432 | Arkansas |
| 050–134 | New York | 433–439 | Louisiana |
| 135–158 | New Jersey | 440–448 | Oklahoma |
| 159–211 | Pennsylvania | 449–467 | Texas |
| 212–220 | Maryland | 468–477 | Minnesota |
| 221–222 | Delaware | 478–483 | Iowa |
| 223–231 | Virginia | 486–500 | Missouri |
| 232–236 | West Virginia | 501–502 | North Dakota |
| 237–246 | North Carolina | 503–504 | South Dakota |
| 247–251 | South Carolina | 505–508 | Nebraska |
| 252–260 | Georgia | 509–515 | Kansas |
| 261–267 | Florida | 516–517 | Montana |
| 268–302 | Ohio | 518–519 | Idaho |
| 303–317 | Indiana | 520 | Wyoming |
| 318–361 | Illinois | 521–524 | Colorado |
| 362–386 | Michigan | 525, 585 | New Mexico |

| | | | |
|---|---|---|---|
| 526–527 | Arizona | 575–576 | Hawaii |
| 528–529 | Utah | 577–579 | District of Columbia |
| 530 | Nevada | 580 | Virgin Islands |
| 531–539 | Washington | 581–585 | Puerto Rico, Guam, |
| 540–544 | Oregon | | American Samoa, |
| 545–573 | California | | Philippine Islands |
| 574 | Alaska | 700–729 | Railroad |

## INVALID SOCIAL SECURITY NUMBERS

1. Three or more leading zeros
2. Ending in four zeros
3. Leading numbers 73 through 79
4. Leading number 6 or 8
5. Leading number of 9 is suspect, very few ever issued

# Appendix
# 2

# Missing and Abused Children Organizations

THE AMERICAN BAR ASSOCIATION
NATIONAL LEGAL RESOURCE CENTER FOR CHILD ADVOCACY AND
PROTECTION
1800 M Street, NW
Washington, DC 20036

CHILD FIND, INC. (NEW YORK ONLY)
P.O. Box 277
New Paltz, NY 12561
800-431-5005

DEE SCOFIELD AWARENESS PROGRAM, INC.
4418 Bay Court Avenue
Tampa, FL 33611

FAMILY AND FRIENDS OF MISSING PERSONS AND VIOLENT
CRIME VICTIMS
P.O. Box 21444
Seattle, WA 98111

FIND THE CHILDREN
11811 W. Olympic Boulevard
Los Angeles, CA 90064
310-477-6721

NATIONAL CENTER ON CHILD ABUSE AND NEGLECT
CHILDREN'S BUREAU/ADMINISTRATION FOR CHILDREN, YOUTH AND
FAMILIES
U.S. Department of Health and Human Services
P.O. Box 1182
Washington, DC 20013

NATIONAL CENTER FOR MISSING AND EXPLOITED CHILDREN
2101 Wilson Boulevard, Suite 550
Arlington, VA 22201
703-235-3900; 800-843-5678

NATIONAL COALITION FOR CHILDREN'S JUSTICE
1214 Evergreen Road
Yardley, PA 19057

NATIONAL COMMITTEE FOR THE PREVENTION OF CHILD ABUSE
Box 2866
Chicago, IL 60607

## RUNAWAY HOT LINES

All across this broad land of ours are organizations, largely in existence through the services of thoughtful and caring volunteers, who are dedicated to helping our troubled youth. Listed below are just a few of what amounts to a crisis telephone network ready with direct or referred information.

Because of the economy offered by no-toll numbers and the availability of cross-references, the location of the hot line sources are not of vital importance, and thus, in some cases, not included here.

CHILD FINDERS OF AMERICA
800-I-AM-LOST

CRISIS CENTER HOT LINE
San Antonio, TX 78228
210-227-4357

CRISIS HOT LINE OF HOUSTON
Houston, TX 77210
713-228-1505

NATIONAL RUNAWAY HOT LINE
Continental United States
800-621-4000

NORTH VIRGINIA HOT LINE
Arlington, VA 22210
703-527-4077

OPERATION PEACE OF MIND
Continental United States
800-231-6946

PHILADELPHIA HOT LINE CONTACT
Philadelphia, PA 19151
215-879-4402

SALVATION ARMY TRANSIENT LODGE HOT LINE
San Antonio, TX 78202
210-226-2291

# Appendix
## 3

# State Medical Boards

If you know your subject's place of birth, true name, and date of birth, here is a possible source that could reveal relatives' names, subject's health history, physical abnormalities, religion, and other details for your profile. Simply write a letter of inquiry.

ALABAMA
P.O. Box 36101
Montgomery, AL 36101

ALASKA
P.O. Box D
Juneau, AK 98811

ARIZONA
3601 W. Camelback Road
Phoenix, AZ 85015

ARKANSAS
P.O. Box 1208
Fort Smith, AR 72902

CALIFORNIA
1426 Howe Avenue
Sacramento, CA 95015

COLORADO
1601 East 19th Avenue
Denver, CO 80218

CONNECTICUT
159 Washington Street
Hartford, CT 06106

DELAWARE
P.O. Box 1401
Dover, DE 19903

DISTRICT OF COLUMBIA
605 G Street NW
Washington, DC 20001

FLORIDA
1940 N. Monroe Street
Tallahassee, FL 32399

GEORGIA
948 Peachtree Street
Atlanta, GA 30309

HAWAII
P.O. Box 3469
Honolulu, HI 32399

IDAHO
407 Bannock Street
Boise, ID 82610

ILLINOIS
320 W. Washington
Springfield, IL 62786

INDIANA
One American Square
Indianapolis, IN 46282

IOWA
1209 W. Court Avenue
Des Moines, IA 50319

KANSAS
235 SW Topeka Boulevard
Topeka, KS 66603

KENTUCKY
400 Sharbon Lane
Louisville, KY 40207

LOUISIANA
830 Union Street
New Orleans, LA 70115

MAINE
State House—Room 137
Augusta, ME 04333

MARYLAND
P.O. Box 22571
Baltimore, MD 21215

MASSACHUSETTS
10 West Street
Boston, MA 02111

MICHIGAN
P.O. Box 22571
Lansing, MI 48908

MINNESOTA
2700 University Avenue
Saint Paul, MN 55114

MISSISSIPPI
2688 Insurance Center
Jackson, MS 33921

MISSOURI
P.O. Box 4
Jefferson City, MO 65102

MONTANA
1424 9th Avenue
Helena, MT 59620

NEBRASKA
P.O. Box 85007
Lincoln, NE 68509

NEVADA
P.O. Box 7238
Reno, NV 89510

NEW HAMPSHIRE
8 Hazen Drive
Concord, NH 03301

NEW JERSEY
228 W. State Street
Trenton, NJ 08608

NEW MEXICO
P.O. Box 20001
Santa Fe, NM 87504

NEW YORK
Empire State Plaza
Albany, NY 12230

NORTH CAROLINA
1313 Navajo Drive
Raleigh, NC 27609

NORTH DAKOTA
418 E. Broadway
Bismarck, ND 59501

OHIO
27 S. High Street
Columbus, OH 43266

OKLAHOMA
P.O. Box 18256
Oklahoma City, OK 73154

OREGON
620 Crown Avenue
Portland, OR 97201

PENNSYLVANIA
P.O. Box 2649
Harrisburg, PA 17105

PUERTO RICO
P.O. Box 9387
Lanturce, PR 00908

RHODE ISLAND
3 Capitol Hill Road
Providence, RI 02904

SOUTH CAROLINA
P.O. Box 12245
Columbia, SC 29211

SOUTH DAKOTA
1329 S. Minnesota Avenue
Sioux Falls, SD 57105

TENNESSEE
283 Plus Park Road
Nashville, TN 37247

TEXAS
P.O. Box 13562
Austin, TX 78711

UTAH
P.O. Box 45802
Salt Lake City, UT 84145

VERMONT
Pavillion Building
Montpelier, VT 05609

VIRGINIA
1601 Rolling Hills Drive
Richmond, VA 23229

WASHINGTON
1300 Quince Street
Olympia, WA 98504

WEST VIRGINIA
101 Dee Drive
Charleston, WV 25311

WISCONSIN
P.O. Box 8935
Madison, WI 53701

WYOMING
2301 Central Avenue
Cheyenne, WY 82001

## YOU'RE WIRED TO UNCLE SAM

For all its enormity, complexities, and tribulations, there is no government on earth more accessible to its constituents. Over one million Americans are government employees. Your subject could be among them. Or it could provide helpful specialized information for your search.

Depending on federal departments remaining after the congressional budget face-off of 1995—and the demands of your search—here is a telephone listing that will reach any one of the departments. Try to match your subject's interests or profession with the department's function. Begin by asking for that department's Personnel Management Office. (All area codes are 202.)

AGRICULTURE
720-8732

COMMERCE
482-2000

DEFENSE
703-545-6700
(See Armed Forces
   locators)

EDUCATION
708-5366

ENERGY
586-5000

HEALTH AND HUMAN
   SERVICES
690-7000

HOUSING AND URBAN
   DEVELOPMENT
708-1422

INTERIOR
208-3100

JUSTICE
514-2000

LABOR
219-6666

STATE
647-4000

TRANSPORTATION
366-4000

TREASURY
622-2000

The *Federal Executive Directory* lists the names, titles, and phone numbers of federal employees. This directory can be found in larger libraries. Or write or call Carroll Publishing Company, 1058 Thomas Jefferson Street NW, Washington, DC 20508; 202-333-8620.

# Appendix
# 4

# Current Adoption Literature

Arty Elbert, author of *Golden Cradle—How the Establishment Works*, suggests you read for general adoption information:

*Adoption: Parenthood Without Pregnancy* by Charlene Canape. A how-to-adopt book touching on the basic issues.

*An Adoptor's Advocate* by Patricia E. Johnston. Humanizing the adoption process, giving you a useful tool to understanding adoption.

*Ideal Adoption: A Comprehensive Guide to Forming an Adoptive Family* by Shirley C. Samuels. A practical look at the adoption process. The reader gets a lot to think about.

*The Adoption Resource Book* by Louis Gilman. A valuable resource for prospective and current adoptive parents. Good material well presented and written.

*The Psychology of Adoption*, edited by David M. Brodzinsky and Marshall D. Schecter. Up-to-date information on research being done, primarily in the area of mental health. Offers insights the reader will find valuable.

*Adopting the Older Child* by Claudia Jewett. A look at the challenge of adding an older child to the family; a look also at critical issues and suggested direction.

*The Handbook for Single Adoptive Parents*, edited by Hope Marindin. Offers counsel on prospective and current single adoptors, and is an excellent general adoption resource.

*Understanding My Child's Korean Origins* by Hyun Sook Han. For anyone planning a Korean adoption.

### LIVING WITH ADOPTION

*How to Raise an Adopted Child* by Judith Schaffer and Christina Lindstrom. These psychotherapists write of situations adoptive parents encounter at various stages.

*Lost and Found: The Adoption Experience* by Betty Jean Lifton. This cogent work argues for the rights of adoptees to search for and know their origins.

*Raising Adopted Children* by Lois Ruskai Melina. Practical and authoritative counsel on special issues. A very helpful and supportive book.

*When Friends Ask about Adoption* by Linda Bothum. Takes on sensitive issues; a good book for friends and relatives of families formed by adoption.

*Adoption Without Fear*, edited by James L. Gritter. First-person narratives help the reader understand what the term "open adoption" really means

*Adoption: A Handful of Hope* by Suzanne Arms contains stories of birth mothers and adoptive families and how open adoption affected their lives.

*Dear Birthmother: Thank You for Our Baby* by Kathleen Silber and Phyllis Speelin. One of the earliest works to deal with open adoption. Contains actual letters exchanged between members of adoptive families, children, and birth parents. If you're considering an open adoption, this book provides food for thought.

*An Open Adoption* by Lincoln Caplan. The process of adopting and the complex relationships that develop among the parties are explored in a true-life case history on an open adoption.

*The Adoption Triangle: Sealed or Open Records? How They Affect Adoptees, Birth Parents, and Adoptive Parents* by Arthur Sorosky, Annette Baran, and Reuben Pannor. A landmark work in adoption first published in 1978, revised to include today's adoption policies. It remains relevant and is highly recommended.

# Appendix
# 5

# The Family History Library and Family History Centers

The Church of Jesus Christ of Latter-day Saints operates the Genealogical Library in Salt Lake City, Utah, and branch genealogical libraries throughout the world. This list includes the branch genealogical libraries in the United States and throughout the world. These libraries have the International Genealogical Index (IGI), the Genealogical Library Catalog (GLC), and other genealogical reference materials. At a branch genealogical library you may also order microfilm from Salt Lake City for a small fee.

Branch genealogical libraries are usually housed in church buildings, and are staffed by volunteers. Each library's schedule varies. Before you visit, contact the library for its hours of operation.

## THE UNITED STATES

ALABAMA

Birmingham 35201
3207 Montevallo Road
205-871-2091

Huntsville 35802
106 Sanders Drive SW
No phone listed

Mobile 36608
5520 Ziegler Boulevard
205-343-9996

Montgomery 36111
3460 Carter Hill Road
205-269-9041

ALASKA

Anchorage 99508
2501 Maplewood Street
907-277-8433

Fairbanks 99701
1500 Cowles Street
907-456-1095

Juneau 99821
5100 Glacier Highway
907-586-2525

Ketchikan 99950
LDS Meetinghouse
907-225-3291

Soldotna 99669
159 Marydale Drive
907-262-4253

Wasilia 99687
Corner of Delwood & Bogard
  Road
907-376-9774

ARIZONA

Cottonwood 83626
127 10th Street
602-634-2349

Flagstaff 86001
625 East Cherry
602-774-8576

Globe 85501
Highway 60
602-425-9570

Holbrook 86025
1600 North 2nd Avenue
602-524-6341

Kingman 86401
3180 Rutherford Drive
602-524-6341

Mesa 85204
464 East First Avenue
602-964-2051

Page 86040
313 Lake Powell Boulevard
602-645-2318

Peoria 85345
12951 North 83rd Avenue
No phone listed

Phoenix 85035
4601 West Encanto Boulevard
602-278-6863

Phoenix 85021
8710 North 3rd Avenue
602-943-3901

Phoenix 85015
3102 North 18th Avenue
602-265-7762

Prescott 86303
1001 Ruth Street
602-778-2311

Safford 85546
501 Catalina Drive
602-778-3194

St. David 85630
Main Street
602-586-4879

St. Johns 85963
35 West Cleveland Street
602-337-2543

Show Low 85901
West Highway 60
602-537-2331

Sierra Vista 85635
Yuka Street
No phone listed

Snowflake 85937
225 West Freeman Avenue
602-536-7430

Tucson 85710
500 South Langley
602-298-0905

Winslow 86047
Kinsley and Lee
602-289-0905

Yuma 85365
4300 West 16th Street
602-782-6364

ARKANSAS

Jacksonville 72076
Highway 67 North
501-982-7967

CALIFORNIA

Anaheim 92801
440 North Loara (Rear)
714-635-2471

Anderson 96007
4075 Riverside Avenue
916-365-8448

Bakersfield 93004
316 A Street
805-325-8907

Barstow 92311
2571 Barstow Road
714-252-4117

Blythe 92226
3rd and Barnard
619-922-4019

Buena Park 90620
7600 Crescent Avenue
714-828-1561

Camarillo 93010
1201 Paseo
805-987-9232

Canyon Country 93010
19513 Drycliff
805-251-5539

Carlsbad 92008
1981 Chestnut Street
619-729-9770

Cerritos 90701
17909 Bloomfield
213-924-3676

Chatsworth 91311
10123 Oakdale Avenue
No phone listed

Chico 95926
2430 Mariposa Avenue
916-343-6641

Covina 91724
656 South Grand Avenue
818-331-7117

El Centro 92244
1280 South 8th Street
619-353-6645

Escondido 92027
1917 East Washington
619-741-8441

Eureka 95501
2734 Dolbeer
707-443-7411

Fairfield 94533
2700 Camrose Drive
707-425-2027

Fresno 93725
6641 East Butler
209-255-4208

Glendale 91206
1130 East Wilson Avenue
818-241-8763

Goleta 93117
478 Cambridge Drive
805-964-8044

Gridley 95948
348 Spruce Street
916-846-3921

Hacienda Heights 91745
16750 Colima Road
818-961-8765

Hemet 92343
425 North Kirby Avenue
714-658-8104

Highland 92343
7000 Central Avenue
714-862-9972

La Crescenta 91214
4550 Raymond Avenue
818-957-0925

Lancaster 93536
3150 West Avenue K Street
805-943-9927

Los Alamitos 90721
4142 Ceritos Avenue
714-821-6914

Los Angeles 90025
Basement of Temple Visitor's
  Center
10741 Santa Monica Boulevard
213-474-2202

Menlo Park 94025
1105 Valparaiso Avenue
415-325-9711

Mission Viejo 92688
23850 Los Alisos Boulevard
714-364-2742

Modesto 95354
731 El Vista Avenue
209-577-9830

Monterey Park 91756
2316 Hillview Avenue
213-726-8145

Napa 94558
Corner of Dry Creek Road and
  Trower Avenue
No phone listed

Needles 92363
El Monte and Lilly Hill Drive
619-326-3363

Newbury Park 91320
35 South Wendy Drive
805-499-7028

Norwalk 90650
15311 South Pioneer Boulevard
213-868-8727

Oakland 94602
4780 Lincoln Avenue
415-531-3905

Orange 92669
674 Yorba Street
714-997-7710

Palmdale 93550
2120 East Avenue R
805-947-1694

Palm Desert 92261
72-960 Park View
619-340-6094

Pasadena 91107
770 North Sierra Madre Villa
213-351-8517

Quincy 95952
Bellamy Lane
916-283-3112

Rancho Palos Verdes 90274
5845 Crestridge
213-541-5644

Redding 96002
3410 Churncreek Road
916-222-4949

Ridgecrest 93555
401 Norma Street
619-375-8100

Riverside 92504
5900 Grand Avenue
714-784-1918

Riverside 92503
4375 Jackson Street
714-687-5542

Sacramento 95821
2745 Eastern Avenue
916-487-2090

San Bernardino 95821
2745 Eastern Avenue
916-487-2090

San Diego 92103
3705 10th Avenue
619-295-0882

San Jose 95112
2175 Santiago Street
408-251-3962

San Luis Obispo 93403
55 Casa Street
805-543-6328

Santa Clara 95051
875 Quince Avenue
408-241-1449

Santa Maria 93454
908 East Sierra Madre Avenue
805-928-4722

Santa Rosa 95403
1725 Peterson Lane
707-525-0399

Seaside 93955
1024 Nocha Buena
408-394-1124

Simi Valley 93063
3979 Township
805-522-2181

Sonora 95370
19481 Hillsdale Drive
No phone listed

Stockton 95207
820 West Brookside Road
209-951-7060

Ukiah 95482
1337 South Dora Street
707-468-5746

Upland 91785
785 North San Antonio
714-985-8821

Ventura 93003
3501 Loma Vista Road
805-643-5607

Victorville 92392
12100 Ridgecrest Road
619-243-5632

Visalia 93277
825 West Tulare Avenue
209-732-3712

Watsonville 95077
255 Holm
408-722-0208

Westminster 92683
10332 Bolsa
714-554-0592

Whittier 90602
7906 South Pickering
213-693-5472

Yuba City 95992
1470 Butte House Road
916-673-0113

Yucaipa 92399
12776 6th Street
No phone listed

COLORADO

Arvada 80004
7080 Independence
303-421-0920

Boulder 80303
701 West South Boulder Road
303-665-4685

Colorado Springs 80907
1054 East Lasalle
303-634-0572

Cortez 81321
1800 East Empire Street
303-565-4372

Craig 81625
11th Finley
No phone listed

Denver 80222
2710 South Monaco Parkway
303-756-6864

Durango 81302
#2 Hill Top Circle
303-259-1061

Ft. Collins 80525
600 East Swallow Drive
303-226-5999

Grand Junction 81506
647 Melody Lane
303-243-2782

Greeley 80631
2207 23rd Avenue
303-353-7941

La Jara 81140
718 Broadway
303-274-4032

Littleton 80123
6705 South Webster
303-973-3727

Littleton 80122
1939 East Easter Avenue
303-798-6461

Montrose 81401
Hillcrest and Stratford
303-249-4281

Northglenn 80233
100 East Malley Drive
303-451-7177

Pueblo 81005
4720 Surfwood
303-564-0793

CONNECTICUT

Bloomfield 06002
1000 Mountain Road
203-649-6547

Madison 06443
275 Warpas
203-245-4986

New Canaan 06840
682 South Avenue
203-966-1305

Quaker Hill 06375
Dunbar Road
203-599-1756

Trumbull 06611
39 Bonnie View
203-374-7444

DELAWARE

Wilmington 19711
143 Dickinson Lane
302-654-1911

DISTRICT OF COLUMBIA

See Kensington, Maryland

FLORIDA

Boca Raton 33486
1530 West Camino Real
305-395-6644

Fort Myers 33901
3105 Broadway
813-936-9831

Gainesville 32605
3745 N.W. 16th Boulevard
904-377-9711

Hialeah 33102
4300 West Fourth Avenue
305-557-9671

Homestead 33030
29600 S.W. 167 Avenue
305-252-2390

Jacksonville 32207
4087 Hendricks Avenue
904-398-3487

Lake Mary 32746
Lake Emma Drive and Greenway
  Boulevard
No phone listed

Marianna 32446
1802 College Street
904-482-8159

Orange Park 32073
461 Blanding Boulevard
904-272-1150

Orlando 32804
45 East Par Avenue
305-898-3841

Panama City 32405
3140 State Avenue
904-785-3601

Pensacola 32504
5773 North 9th Avenue
904-476-0183

Rockledge 32955
1801 Fisk Boulevard
305-636-2431

St. Petersburg 33702
570 62nd Avenue North
813-525-9351

Tallahassee 32304
312 Stadium Drive
904-224-6431

Tampa 33637
4106 Fletcher Avenue
813-971-2869

Winter Haven 33880
1958 9th Street SE
813-299-9431

GEORGIA

Columbus 31907
Reese Road
No phone listed

Dunwoody 30338
1155 Mt. Vernon Highway
404-393-4329

Macon 31206
1624 Williamson Road
912-788-3064

Marietta 30060
1578 Cunningham Road South
No phone listed

Savannah 31406
613 Montgomery Cross Road
912-927-6543

Valdosta 31602
1307 West Alden Avenue
912-242-2300

HAWAII

Hilo 96721
1373 Kilauea Avenue
808-935-0711

Honolulu 96826
1560 South Beretania Street
808-955-8910

Honolulu 96819
1733 Beckley Street
808-841-4118

Kaneohe 96744
46-117 Halaulani Street
808-247-3134

Kona 96747
Kalani Road
808-329-5054

Laie 96762
55-600 Naniloa Loop
808-293-2133

Mililani 96789
95-186 Wainaku Place
No phone listed

IDAHO

Arco 83213
Country Road
208-527-8900

Blackfoot 83221
Old Seminary Road
208-785-5022

Blackfoot 83221
101 North 900 West
208-684-3784

Boise 83702
325 West State Street
208-334-2305

Boise 83709
12040 West Amity Road
208-362-5847

Burley 83318
224 East 14th Street
208-678-7286

Caldwell 83605
3015 South Kimball
208-459-2531

Coeur d'Alene 83814
2801 North 4th
208-765-0150

Driggs 83422
221 North 1st East
208-354-2253

Emmett 83617
980 West Central Road
208-365-4112

Firth 83236
East Center Street
208-346-6282

Idaho Falls 83401
1155 1st Street
208-524-5291

Idaho Falls 83406
3000 Central Avenue
208-529-4087

Idaho Falls 83401
1860 Kearny
208-529-9805

Iona 83427
Iona North Road and US 26
No phone listed

Lewiston 83501
9th and Preston
208-743-9744

Malad 83252
312 West 400 North
208-766-2332

Meridian 83624
Shamrock and McMillan Road
208-376-4052

Montpelier 83254
Bear Lake Country Library
138 North 6th Street
208-847-0340

Nampa 83686
143 Central Canyon
208-467-5827

Pocatello 83201
156¹/₂ South 6th Center
208-232-9262

Rexburg 83440
Ricks College Library
208-232-9262

Salmon 83467
Stake Center
208-756-3514

Sandpoint 83862
433 South Boyer
208-263-8721

Shelly 83274
325 East Locust
208-357-7831

Soda Springs 83276
281 East Hooper Avenue
208-547-2237

Twin Falls 83303
401 Maurice Street North
208-733-8073

Weiser 83672
306 East Main Street
208-549-1575

ILLINOIS

Carbondale 62901
Old Route 13
618-549-3034

Champaign 61820
604 West Windsor Road
217-352-8063

Chicago Heights 60620
402 Longwood Drive
312-754-2525

Naperville 60540
25 West 341 Ridgeland Road
312-357-0211

Nauvoo 62354
Durphy Street
217-453-6347

Peoria 61615
3700 West Reservoir Boulevard
309-682-4073

Rockford 61107
620 North Alpine Road
815-399-5448

Schaumburg 60194
1320 West Schaumburg Road
312-882-9889

Wilmette 60091
2801 Lake Avenue
312-251-9818

INDIANA

Bloomington 47402
2411 East 2nd Street
812-332-9786

Evansville 47708
519 East Almstead
812-423-9832

Ft. Wayne 46835
5401 St. Joe Road
219-485-9581

Indianapolis 46227
900 East Stop 11 Road
317-888-6002

New Albany 47150
1534 Slate Run Road
No phone listed

Noblesville 46060
777 Sanblest Boulevard
317-849-6086

South Bend 46619
3050 Edison Road
219-233-6501

Terre Haute 47804
1845 North Center
812-234-0269

IOWA

Ames 50010
2524 Hoover
515-232-3434

Cedar Rapids 52403
4300 Trailridge Road SE
319-386-7547

Davenport 52806
4929 Wisconsin Avenue
319-386-7547

Sioux City 51101
1201 West Clifton
712-255-9686

West Des Moines 50265
3301 Ashworth Road
515-225-0416

KANSAS

Dodge City 67801
2506 6th Avenue
316-225-6540

Olathe 66062
15916 West 143rd Street
913-829-1775

Topeka 66611
3611 SW Jewell
913-266-7503

Wichita 67206
7011 East 13th Street
316-683-2951

KENTUCKY

Hopkinsville 42240
1118 Pin Oak Drive
No phone listed

Lexington 40502
1789 Tates Creek Park
606-269-2722

Louisville 40222
1000 Hurstborne Lane
502-426-5317

Paducah 42001
320 Birch Street
502-442-5317

LOUISIANA

Alexandria 71303
611 Versailles Street
318-448-1842

Baton Rouge 70805
5686 Winbourne Avenue
504-357-8385

Metairie 70112
5025 Cleveland Place
504-885-3936

Monroe 71201
909 North 33rd Street
318-387-3793

Shreveport 71105
200 Carroll Street
318-868-5169

MAINE

Bangor 04401
639 Grandview Avenue
207-942-7677

Cape Elizabeth 04104
29 Ocean House Road
207-799-7018

Caribou 04736
46 Hardison
No phone listed

Farmingdale 04345
Hasson Street
207-582-4686

MARYLAND

Ellicott City 21043
4100 St. Johns Lane
301-465-1642

Frederick 21701
199 North Place
No phone listed

Kensington 20895
10000 Stoneybrook Drive
301-587-0144

Lutherville 21093
1400 Dulaney Valley Road
301-256-5890

MASSACHUSETTS

Foxboro 02035
76 Main
617-543-5284

Weston 02138
150 Brown Street
617-235-9892

MICHIGAN

Ann Arbor 48104
914 Hill Street
313-995-0211

Bloomfield Hills 48302
425 North Woodward Avenue
313-647-5671

East Lansing 48906
431 East Saginaw Street
527-332-2932

Grand Blanc 48439
4285 McCandlish Road
616-949-0070

Grand Rapids 49506
3181 Bradford NE
616-949-0070

Kalamazoo 49007
1112 North Drake Road
616-342-1906

Marquette 49855
Cherry Creek Road
906-249-1511

Midland 48640
1700 West Sugnet Road
517-631-1120

Westland 48185
7575 North Hix Road
313-459-4570

MINNESOTA

Duluth 55811
521 Upham Road
218-722-9508

Minneapolis 55422
2801 North Douglas Drive
612-544-2479

Rochester 55904
1002 Southeast 16th Street
507-282-2382

St. Cloud 56302
1420 29th Avenue North
612-252-4355

St. Paul 55119
2200 North Haley
612-770-3213

MISSISSIPPI

Clinton 39202
1301 Pinchaven Road
601-924-2537

Columbus 39701
708 Airline Road
No phone listed

Gulfport 39507
Klein Road at David
601-832-0195

Hattiesburg 39401
US 11 South
601-544-9238

MISSOURI

Columbia 65202
904 Old Highway 36 South
314-443-1024

Frontenac 63141
10445 Clayton Road
314-993-2328

Independence 64050
705 West Walnut
816-461-0245

Joplin 64804
22nd and Indiana
417-623-6506

Kansas City 64131
8144 Holmes
816-444-3444

Liberty 64068
1120 Clayview Drive
816-781-8295

Springfield 65804
3325 East Bennett
417-887-8229

MONTANA

Billings 59102
1711 Sixth Street West
406-245-8112

Billings 59105
1000 Wicks Lane
406-259-3348

Bozeman 58715
2195 Coutler Drive
406-586-3880

Butte 59701
3400 East 4 Mile Road
406-494-9909

Great Falls 59404
1401 9th Street NW
406-453-4280

Helena 59601
1610 East 6th Avenue
406-442-1558

Kalispell 59903
Buffalo Hill and Bountiful Drive
406-543-5446

Missoula 59801
3201 Bancroft Street
406-543-6148

Stevensville 59870
Eastside Highway and Middle-
    Burnt Fork Road
No phone listed

NEBRASKA

Grand Island 68801
212 West 22nd Street
308-382-9418

Lincoln 68516
3100 Old Cheney Road
402-423-4561

Omaha 68144
11027 Martha Street
402-393-7641

NEVADA

Elko 89801
1651 College Parkway
702-738-4565

Ely 89301
900 Avenue East
702-289-3575

Fallon 89406
750 West Richards Street
702-423-2094

Las Vegas 89104
509 South Ninth Street
702-382-9695

Logandale 89021
Highway 12
702-393-3594

Reno 89501
Washoe Public Library
301 South Center
702-785-4530

Sparks 89431
2955 North Rock Boulevard
702-359-5834

Winnemucca 89445
111 West MacArthur Avenue
702-623-4413

NEW HAMPSHIRE

Concord 03301
90 Clinton Street
603-224-3061

Nashua 03060
110 Concord Street
603-880-7371

NEW JERSEY

East Brunswick 08816
303 Dunham's Corner Road
201-254-1480

Moorestown 08057
Bridgeboro Road
609-234-9639

Morristown 07960
283 James Street
201-539-5362

NEW MEXICO

Albuquerque 87107
1100 Montano Road NW
505-345-0406

Albuquerque 87110
5709 Haines Avenue NE
505-226-4867

Carlsbad 88221
1200 West Church and Oak Street
505-885-1368

Farmington 87499
400 West Apache
505-325-5813

Gallup 87301
601 Susan Avenue
No phone listed

Grants 87020
1010 Bondad
505-287-2548

Las Cruces 88001
1015 Telsher Boulevard
505-522-2300

Santa Fe 87505
410 Rodeo Road
No phone listed

NEW YORK

Jamestown 14702
851 Forest Avenue
716-487-0830

Loudonville 12211
411 Loudon Road
518-462-3687

New York City 10023
2 Lincoln Square (3rd floor)
Broadway at 65th Street
212-799-2660

Plainview 11803
160 Washington Avenue
516-433-0122

Pittsford 14534
460 Kreag Road
716-248-9930

Syracuse 13214
801 East Colvin Street
315-478-8484

Vestal 13850
305 Murray Hill Road
607-798-7424

Williamsville 14204
1424 Maple Road
716-688-9759

NORTH CAROLINA

Arden 28704
Highway 25A, Rosscraggon Road
No phone listed

Charlotte 28205
3020 Hilliard Drive
704-535-0238

Fayetteville 28303
3200 Scotty Hill Road
919-864-2080

Greensboro 27410
3719 Pinetop Road
919-864-6539

Hickory 28601
Highway 127 North
704-328-4077

Kingston 28501
3006 Carey Road
919-522-4671

Raleigh 27609
5100 Six Forks Road
919-781-1662

Wilmington 27604
514 South College Road
919-395-4456

Winston-Salem 27103
4760 Westchester Drive
919-765-7231

NORTH DAKOTA

Bismarck 58501
1500 Country West Road
701-225-4590

Fargo 58103
2502 17th Avenue South
701-293-9362

Minot 58701
2025 9th Street NW
701-838-4486

OHIO

Akron 44313
735 North Revere Road
216-864-0203

Cincinnati 45242
Cornell and Snider
513-489-9957

Cincinnati 45212
5505 Bosworth Place
515-531-5624

Columbus 43209
3648 Leib Street
614-451-0483

Dayton 45426
1500 Shiloh Springs Road
513-854-4566

Fairborn 45324
3080 Bell Drive
513-878-9551

Kirkland 44094
Chillicothe Road
216-256-8808

Reynoldsburg 43068
2135 Baldwin Road
614-866-7686

Toledo 43619
1545 East Gate
419-382-0262

Westlake 44145
25000 Westwood Road
216-777-1518

OKLAHOMA

Lawton 73507
923 Hilltop Drive
405-355-9946

Muskogee 74403
3008 East Hancock Road
918-687-8861

Norman 73071
Imhoff Road and Highway 9
405-364-8337

Oklahoma City 73132
5020 Northwest 63rd
405-721-8455

Stillwater 74075
1720 East Virginia
405-372-8569

Tulsa 74128
12110 East 7th Street
918-437-5690

OREGON

Beaverton 97007
4195 Southwest 99th
503-644-7782

Bend 97701
1260 Thompson Drive
503-389-3559

Central Point 97502
2305 Taylor Road
No phone listed

Coos Bay 97402
3950 Sherman Avenue
503-269-9037

Corvallis 97330
4141 NW Harrison
503-752-2256

Eugene 97412
3550 West 18th Street
503-343-3741

Grants Pass 97527
1969 Williams Highway
502-479-7644

Gresham 97030
3500 Southeast 182nd Street
503-665-1524

Hermiston 97838
850 Southwest 11th
503-567-3445

Hillsboro 97124
2200 Northeast Jackson School
  Road
503-640-4658

Klamath Falls 97603
McClellan Drive at Aiva
502-884-2133

La Grande 97850
2504 North Fir
503-963-5003

Lake Oswego 97034
1271 Overlook Drive
503-638-8486

McMinnville 97128
1645 Northwest Baker Creek Road
503-434-5681

Medford 97504
2900 Juanipero Way
503-773-3363

Nyssa 97913
West Alberta Avenue
503-773-3363

Oregon City 97045
14340 South Donovan Road
503-657-9584

Portland 97214
2931 Southeast Harrison
503-235-9090

Portland 97220
2215 Northeast 106th Street
503-252-1031

Prineville 97754
Idlewood and South 2nd
503-447-1488

Roseburg 97470
1864 Northwest Calkins Road
503-672-4108

Salem 97301
862 45th NE
403-371-0453

Salem 97302
4550 Lone Oak SE
503-363-0374

Salem 97303
1395 Lockhaven Drive NE
503-390-2095

Sandy 97055
Gresham Oregon South
16317 SE Bluff
No phone listed

The Dalles 97058
15th and Oregon
503-296-4301

PENNSYLVANIA

Broomall 19008
721 Paxon Hollow Road
215-356-8507

Clarks Summit 18411
Leach Hill and Griffin Pond Roads
717-587-5123

Erie 16505
1101 South Hill Road
814-866-3611

Knox 16232
Clarion Meetinghouse
814-797-1287

Pittsburgh 15220
46 School Street
412-921-2115

Reading 19606
3344 Reading Crest Avenue
215-929-0235

State College 16802
842 Whitehall Road
814-238-4560

York 17315
2100 Hollywood Drive
717-854-9331

RHODE ISLAND

See listing for Quaker Hill,
   Connecticut

SOUTH CAROLINA

Charleston 29407
1310 Sam Rittenburg (Highway 7)
803-766-6017

Columbia 29209
4440 Ft. Jackson Boulevard
803-782-7141

Florence 29502
1620 Maldin Drive
803-662-9482

Greenville 29611
Farr's Bridge Road
No phone listed

SOUTH DAKOTA

Rapid City 57702
2822 Canyon Lake Drive
605-341-8572

Rosebud 57570
LDS Meetinghouse
No phone listed

Sioux Falls 57106
3900 South Fairhall Avenue
605-361-1070

TENNESSEE

Chattanooga 37411
1019 North Moore Road
615-892-7632

Kingsport 37660
100 Cannongate Road
615-245-2321

Knoxville 37919
400 Kendall Road
615-690-4041

Madison 37115
107 Twin Hills Drive
615-859-6926

Memphis 38119
8150 Walnut Grove Road
901-754-2545

TEXAS

Abilene 79603
3325 North 12th Street
915-673-8836

Amarillo 79109
5401 Bell Street
806-355-4796

Austin 78753
1000 East Rutherford
512-837-3626

Bryan 77802
1200 Barak Lane
409-846-7929

Corpus Christi 78414
6750 Woodridge Road
512-993-2970

Dallas 75205
10701 Highlands Drive
No phone listed

Denton 76201
1801 Malone
817-387-3065

Duncanville 75137
1019 Big Stone Gap
214-709-0066

El Paso 79903
3651 Douglas Avenue
915-565-9711

Fort Worth 76134
5001 Altamesa Boulevard
817-292-8393

Friendswood 77546
505 Deseret
713-996-9346

Harlingen 78550
2320 Haine Drive
No phone listed

Houston 77070
10555 Mills Road
713-890-7434

Houston 77057
1101 Bering Drive
713-785-2105

Houston 77017
3000 Broadway
713-893-5381

Houston 77090
16331 Hafer Road
713-893-5381

Killeen 77449
1410 South 2nd Street
817-526-2918

Kingwood 77339
4021 Deerbrook
713-360-1363

Longview 75605
1700 Blueridge Parkway
214-759-7911

Lubbock 79413
3211 58th Street
806-792-5040

McAllen 78501
2nd Street La Vista
512-682-0051

Odessa 79761
2011 North Washington
915-332-9221

Orange 77631
6108 Hazelwood
No phone listed

Plano 75075
2700 Roundrock
214-867-6479

Port Arthur 77642
3939 Turtle Creek Drive
409-722-4997

Richland Hills 75239
4401 Northeast Loop 820
817-284-4472

San Antonio 78228
2103 St. Cloud
512-736-2904

UTAH

Altamont 84001
Main Street
801-438-5282

Beaver 84713
210 North Main
801-438-5262

Blanding 84511
225 East 2nd North
801-678-2024

Bountiful 84010
165 South 1000 East
801-292-2650

Brigham City 84302
10 South 4th East
801-723-5995

Castledale 84513
Stake Center
801-748-2555

Cedar City 84720
370 South 200 East
801-586-2296

Delta 84624
52 North 100 West
801-864-3312

Duchesne 84021
Stake Center
801-738-5371

Ferron 84523
400 West 530 South
No phone listed

Fillmore 84631
21 South 300 West
801-743-6623

Garland 84321
131 East 1500 South
801-257-7015

Heber City 84032
781 South 200 East
801-654-2760

Hurricane 84737
37 South 200 West
801-635-2174

Kanab 84741
218 East 200 North
801-644-5973

Lehi 84043
200 North Center Street
801-644-5973

Loa 84747
20 South 100 West
801-836-2651

Logan 84321
50 North Main
801-836-2651 or 801-752-0546

Moroni 84646
300 North Center Street
801-436-8497

Mt. Pleasant 84647
295 South State
No phone listed

Nephi 84648
351 North 100 West
801-623-1378

Ogden 84401
539 24th Street
801-393-5248

Parowan 84761
87 West Center Street
801-477-8077

Price 84501
85 East Fourth North
801-637-2071

Provo 84602
Brigham Young University
4386 HBL Library
801-378-6200

Richfield 84701
91 South 200 West
801-896-8057

Roosevelt 84060
447 East Lagoon Street
801-722-3794

St. George 84070
58 East 9639 South
No phone listed

Sandy 84106
1700 East 775 South
801-673-4591

Sandy 84070
1700 East 9639 South
No phone listed

Santaquin 84655
90 South 200 East
801-754-3725

South Jordan 84065
2450 West 10400 South
801-254-0121

Springville 84663
900 East 200 North
801-489-9966

Tropic 84776
LDS Meetinghouse
No phone listed

Vernal 84008
613 West 2nd South
801-789-2618

Wellington 84542
Stake Center
No phone listed

Wendover 84083
269 B Street
801-665-2220

VERMONT

Berlin-Montpelier 05602
James Parker, Sr.
RR #1, Box 2560
No phone listed

VIRGINIA

Annandale 22003
3900 Howard Street
703-256-5518

Charlottesville 22901
Hydraulic Road
804-973-6607

Dale City 22193
3000 Dale Boulevard
702-670-5977

Newport News 23602
901 Denbigh Boulevard
804-874-2335

Oakton 22124
2719 Hunter Mill Road
703-281-1836

Richmond 23226
5600 Monument Avenue
804-288-8134

Salem 24153
6311 Watburn Drive
703-366-6726

Virginia Beach 23462
4760 Princess Anne Road
804-467-3302

WASHINGTON

Bellevue 98004
10675 Northeast 20th Street
206-454-2690

Bremerton 98315
2225 Perry Avenue
206-479-9370

Centralia 98531
2801 Mt. Vista Road
206-736-5476

Edmonds 98020
7309 228th Street SW
206-774-0933

Elma 98541
702 East Main
206-482-5982

Everett 98201
9505 19th Avenue SE
206-337-0457

Federal Way 98003
34815 3rd Avenue S
206-874-3803

Ferndale 98248
5800 Northwest Avenue
206-384-6188

Kennewick 99336
515 South Union
No phone listed

Lake Stevens 98258
131 101st Avenue SE
206-335-0754

Longview 98632
1721 30th Avenue
206-425-8409

Moses Lake 98837
1515 Division
509-765-8711

Mount Vernon 98273
18th and Hazel
No phone listed

North Bend 98045
527 Mt. Si Boulevard
No phone listed

Olympia 98506
Puget and Yew Street
209-943-7055

Othello 99327
12th and Rainier
509-488-6437

Pasco 99301
2108 Road 24
509-488-6437

Pullman 99163
NE 1055 Orchard Drive
No phone listed

Quincy 98848
1102 2nd Avenue SE
No phone listed

Richland 99352
1314 Goethals
509-946-6637

Seattle 98166
142nd SE and Ambaum
   Boulevard SW
206-246-7864

Seattle 98105
5701 8th NE
206-522-1233

Silverdale 98383
Nels Nelson Road
No phone listed

Spokane 99206
N. 919 Pines Road
509-926-0551

Sumner 98390
512 Valley Avenue
206-863-3383

Tacoma 98465
South 12th and Pearl Street
206-564-1103

Vancouver 98684
10509 Southeast 5th Street
206-256-7235

Walla Walla 99362
1821 South 2nd Street
509-525-1121

Wenatchee 98801
667 10th NE
509-884-3285

Yakima 98902
705 South 38th Avenue
No phone listed

WEST VIRGINIA

Fairmont 26555
Route 73 South
304-363-0116

Huntington 25705
5640 Shawnee Drive
304-736-9072

WISCONSIN

Hales Corner 53130
9600 West Grange Avenue
414-425-4182

Madison 53705
1711 University Avenue
608-238-1071

Shawano 54166
910 East Zingler
715-526-2946

WYOMING

Afton 83110
347 Jefferson
307-886-3526

Casper 82604
Corner of Fox and 45th
307-234-3326

Cheyenne 82001
Wyoming County Library
2800 Central Avenue
307-634-3561

Cody 82001
1407 Heart Mountain
307-634-3561

Evanston 82930
1224 Morse Lee Street
No phone listed

Gillette 82716
1500 O'Hara Street
307-686-2077

Green River 82935
120 Shoshone Avenue
307-875-3972

Jackson 83001
520 East Broadway
307-733-6337

Kemmerer 83101
Antelope 3rd West Avenue
307-877-6502

Laramie 82070
1219 Grande Avenue
No phone listed

Lovell 82431
50 West Main
307-548-2963

Rawlins 82301
117 West Kendrick Street
No phone listed

Riverton 82501
923 Garfield
301-332-3666

Rock Springs 82901
2055 Edgar Street
307-362-8062

Sheridan 82801
2051 Colonial Drive
307-674-9904

Urie 82937
Lyman Stake Center
307-786-4559

Worland 82401
500 Sagebrush Drive
307-347-8958

## CANADA

### ALBERTA

Calgary
910 70 Avenue SW
403-244-5910

Cardston
846 1st Avenue West
403-653-3288

Edmonton Alberta
9010 85 Street NW
403-469-6460

Grande Prairie
11212 102 Street
403-532-3609

Lethbridge
2410 28th Street S
403-328-0206

Raymond-Magrath
2 North 200 West
403-752-3833

Red Deer
3002 47 Avenue
403-342-1508

Taber
50th Avenue and 47th Street
403-223-2243

### BRITISH COLUMBIA

Burnaby
5280 Kincaid
604-299-8656

Cranbrook
2210 2nd Street N
No phone listed

Fort St. John
11412 100 Street
604-785-4354

Kamloops
2165 Parkcrest Avenue
No phone listed

Kelowna
Glenmore and Ivans Street
604-762-0588

Prince George
4108 5th Avenue
604-563-1490

Vancouver
See Burnaby

Vernon
See Kelowna

Victoria
701 Mann Avenue
604-479-3631

MANITOBA

Winnipeg
700 London Street
204-668-0091

NEW BRUNSWICK

St. John
177 Manchester Avenue
506-672-0864

NOVA SCOTIA

Dartmouth
44 Cumberland Drive
902-462-0628

ONTARIO

Glenburnie
Glenburnie and Battersea Road
613-544-0221

Hamilton
701 Stonechurch Road E
416-385-5009

Kingston
See Glenburnie

London
1139 Riverside Drive
519-473-2421

Northland
See Timmins

Oshawa
Corner of Thornton Road and
   Rossland Road
416-728-3151

Ottawa
1017 Prince of Wales Drive
613-224-2231

Sault Ste. Marie
No address yet
No phone listed

Thunder Bay
2255 Ponderosa Drive
807-939-1451

Timmins
500 Toke Street
705-267-6400

Toronto
95 Melbert Street
416-621-4607

Toronto
See also Oshawa

QUEBEC

Montreal (French)
470 Gilford
514-849-7145

Montreal (English)
6666 Terre Bonne Street
514-489-9138

SASKATCHEWAN

Saskatoon
1427 10th Street E
No phone listed

## BRITISH ISLES

ENGLAND

Ashton
See Rochdale

Billingham
The Linkway
Billingham, Cleveland
Tel. 0642-563-162

Birmingham
see Sutton Coldfield

Bristol
721 Wells Road
Whitchurch, Bristol, Avon
Tel. 0272-560-106

Cambridge
670 Cherry Hinton Road
Cambridge, Cambridgeshire
Tel. 0223-247-010

Cheltenham
Thirlestaine Road
Cheltenham, Gloucestershire
Tel. 0242-523-433

Chester
50 Cliftone Drive
Blacone, Chester, Cheshire
Tel. 0244-390-796

Crawley
Old Horsham Road
Crawley, West Sussex
Tel. 0293-516-151

Hornchurch
64 Butts Green Road
Hornchurch, Essex
Tel. 0424-58412

Huddersfield
12 Halifax Road
Birchencliffe, Huddersfield,
    West Yorkshire
Tel. 0484-20352

Hull
Springfield Way
Anlaby, Near Hull, Humberside
Tel. 0482-572-623

Hyde Park
See London

Ipswich
42 Sidegate Lane West
Ipswich, Suffolk
Tel. 0473-723182

Leeds
Vesper Road
Leeds, West Yorkshire
Tel. 0532-585297

Leicester
See Loughborough

Lichfield
Purcell Avenue
Lichfield, Staffordshire
Tel. 0543-262621

Liverpool
4 Mill Bank
Liverpool, Merseyside
Tel. 051-228-0433

London
64/68 Exhibition Road
South Kensington, London
Tel. 01-589-8561

Loughborough
Thorpe Hill, Alan Moss Road
Loughborough, Leicestershire
Tel. 0509-214-991

Luton
Cutenhoe Road and London Road
Luton, Bedfordshire
Tel. 0582-22242

Maidstone
London Road
Maidstone, Kent
Tel. 0622-57811

Manchester
Altrincham Road
Wythenshawe, Greater
    Manchester
Tel. 061-902-9279

Newcastle-under-Lyme
The Brampton
Newcastle-under-Lyme,
    Staffordshire
Tel. 0782-620-653

Northampton
137 Harlestone Road
Northampton, Northamptonshire
Tel. 0604-51348

Norwich
19 Greenways
Eaton, Norwich, Norfolk
Tel. 0603-52440

Nottingham
Hempshill Lane
Bulwell, Nottinghamshire
Tel. 0602-274-194

Plymouth
Hartley Chapel, Mannamead Road
Plymouth, Devon
Tel. 0752-668-998

Poole
8 Mount Road
Parkstone
Poole, Dorset
Tel. 0202-730-646

Preston
See Rawtenstall

Rawtenstall
Haslingden Road
Rawtenstall
Rossendale, Lancashire
Tel. 0706-213-460

Reading
280 The Meadway
Tilehurst
Reading, Berkshire
Tel. 0734-427-524

Rochdale
Tweedale Street
Rochdale, Greater Manchester
Tel. 0706-526-292

Romford
See Hornchurch

St. Albans
See Luton

Sheffield
Wheel Lane, Grenoside
Sheffield, South Yorkshire
Tel. 0742-453231

Southampton
Chetwynd Road
Bassett, Southampton, Hampshire
Tel. 0703-767-476

Staines
41 Kingston Road
Staines, Middlesex
Tel. 0784-50709

Sunderland
Linden Road off Queen Alexandra
   Road
Sunderland, Tyne & Wear
Tel. 091-528-5787

Sutton Coldfield
185 Penns Lane
Sutton Coldfield
Birmingham, West Midlands
Tel. 021-384-2028

Wandsworth
484 London Road
Mitcham, London
Tel. 769-0180

York
West Bank
Acomb
York, North Yorkshire
Tel. 0904-798-185

CHANNEL ISLANDS

St. Helier
Rue De La Vallee
St. Mary
Jersey, C. I.
Tel. 0534-82171

IRELAND

Dublin
The Willows, Finglas Road
Dublin, County Dublin
Tel. 010-353-4625609

NORTHERN IRELAND

Belfast
401 Holywood Road
Belfast, County Antrim
Tel. 0232-768250

SCOTLAND

Aberdeen
North Anderson Drive
Aberdeen, Grampian
Tel. 0224-692-206

Dundee
Bingham Terrace
Dundee, Tayside
Tel. 0382-4512474

Glasgow
35 Julian Avenue
Glasgow, Strathclyde
Tel. 041-357-1024

Paisley
Campbell Street
Johnstone, Strathclyde
Tel. 0505-20886

WALES

Cardiff
Heol Y Deri
Rhiwbina
Cardiff, South Glamorgan
Tel. 0222-620205

Merthyr Tydfil
Nant-y-gwenith Street
George Town, Merthyr Tydfil,
   Mid Glamorgan
Tel. 0685-2455

## CONTINENTAL EUROPE

### AUSTRIA

Vienna
Boecklinser 55
Tel. 0222-735-649

### BELGIUM

Brussels
See Grimbergen

Grimbergen
Strombeeklinde, 110
No phone listed

### DENMARK

Copenhagen
See Frederiksberg

Frederiksberg
Priorvej 12
No phone listed

### FINLAND

Oulu
Nokelantie 38
Tel. 981-335714

### FRANCE

Nancy
See Schiltigheim

Nice
5 Avenue Thérèse
Tel. 9381-0669

Paris
See Versailles

Schiltigheim
100 Route de General de Gaulle
Tel. 88-83-62-65

Strasbourg
See Schiltigheim

Versailles
5 Rond Point de l'Alliance
Tel. 3-954-82-78

### GERMANY

Berlin
Klingelhoefer Strasse 24
Tel. 030-262-1089

Düsseldorf
See Wuppertal

Frankfurt
Eckenheimer Landstrasse 264
Tel. 069-546005

Hamburg
Wartenau 20
Tel. 040-2504573

Hamburg
See also Neumunster

Hannover
See Stadthagen

Kaiserslautern
Lauterstrasse 1
Tel. 0631-79588

Mannheim
See Nussloch

Munich
Rueckerstrasse 2
Tel. 089-535176

Neumunster
Kieler Strasse 333
Tel. 04321-38548

Nuremberg
Kesslerplatz 8
No phone listed

Nussloch
Menzelstrasse 9
No phone listed

Stadthagen
Jahnstrasse/Ecke Schactstrasse
Tel. 05721-74459

Stuttgart
Birkenwaldstrasse 46
Tel. 0711-224871

Wuppertal
Martin-Luther-Strasse 6
Tel. 0202-89158

## ICELAND

Reykjavik
Skolavordustig 46
Tel. 91-28730

## ITALY

Milan
Casa Riunione
Tel. 02-28-22-58

## NETHERLANDS

Apeldoorn
Boerhavestr/Edisonlaan
Tel. 055-217516

Utrecht
See Apeldoorn

## SWEDEN

Stockholm
See Västerhaninge

Västerhaninge
Tempelvägen 3
Tel. 0750-28520

## SWITZERLAND

Bern
See Ersigen

Cointrin
32 Avenue Louis Casai
Tel. 22-98-63-57

Ersigen
Sandrutiweg 11
No phone listed

Geneva
See Cointrin

Zurich
Herbstweg 120
Tel. 0041-1413598

SPAIN
Barcelona
No address or phone yet

Madrid
No address or phone yet


## AUSTRALIA, NEW ZEALAND, AND SOUTH PACIFIC

AUSTRALIA

*Australian Capital Territory*

Canberra
101-105 Wattle Street
O'Connor
Tel. 062-47-5876

*New South Wales*

Greenwich
55 Greenwich Road
Tel. 02-43-10-05

Hebersham
64-66 Pringle Road
Tel. 02-625-3365

Lismore
Glen Court
Goonellabah
Tel. 28-1330

Mortdale
74 Walter Street
Tel. 02-570-6453

Newcastle
5 Nalya Close
Charlestown
Tel. 049-43-0122

Orange
Frost Street
Tel. 063-62-9192

Parramatta
169 Pennant Street
North Parramatta
Tel. 02-630-1931

Sydney
See Greenwich, Hebersham,
    Mortdale, and Parramatta

Tamworth
Ridge Street
No phone listed

*Northern Territory*

Darwin
Trower and Sabine
Tel. 089-85-1980

*Queensland*

Brisbane
Underwood Road
Eight Mile Plains
Tel. 07-341-4485

200 River Terrace
Kangaroo Point
Tel. 07-391-7585

Bundaberg
Corner of Bingera and Woongarra
    Streets
Tel. 071-73-1701

Cairns
Oxley Street
Edgehill
Tel. 070-53-1503

Ipswich
Corner Haig and Hunter Street
Tel. 07-201-7823

Rockhampton
Talbot Street
North Rockhampton
Tel. 079-28-3434

Townsville
188 Fulham Road
Tel. 077-79-3028

*South Australia*

Adelaide
Cutting Road
Marion
Tel. 08-276-7849

Von Braun Crescent
Modbury North
Tel. 08-264-7010

Mount Gambier
Brigalow Crescent
No phone listed

Port Pirie
Dunn Street
Tel. 086-32-1964

Whyalla
Jenkins Avenue
Tel. 086-45-7647

*Tasmania*

Hobart
15 Elmsleigh Road
Moonah
Tel. 002-72-9529

*Victoria*

Geelong
Corner Wolsely Grove and
   Eagleview Pdg.
Tel. 052-78-1691

Melbourne
285 Heidelberg Road
Northcote
Tel. 03-481-7079

Cathies Lane and Pumps Road
Wantirna
Tel. 03-801-9746

Mildura
Deakin Avenue
Tel. 050-23-3576

Traralgon
Kosciusko Street
No phone listed

Wangaratta
45 Garnett Avenue
Tel. 057-21-5486

*Western Australia*

Kalgoorlie
118 Ega Street
No phone listed

Perth
308 Preston Point Road
Atadale
Tel. 09-330-3750

163 Wordsworth Avenue
Yokine
Tel. 09-275-2608

COOK ISLANDS

Raratonga
No address listed
No phone listed

FIJI

Suva
15 Helsen Street, Samabula
Tel. 679-383-284

NEW ZEALAND

Auckland
2 Rowandale Road
Manurewa
Tel. 267-5479

147 Pah Road
Mt. Roskill
Tel. 659-669

9 Tahoroto Road
Takapuna
Tel. 480-5828

54 Rua Road
Glen Eden
Tel. 818-3656

Christchurch
25 Fendalton Road
Tel. 556-874

Dunedin
Fenton Crescent
Tel. 879539

Gisborne
290 Stout Street
Tel. 798941

Hamilton
Sandwich Road
Tel. 491758

Hastings
Heretaunga Street
Tel. 83717

Kaikohe
Hongi Street
No phone listed

Nelson
Nayland Road Stoke
Tel. 79507

New Plymouth
No address listed
No phone listed

Palmerston North
29 Pitama Road
Tel. 72011

Rotorua
10 Rimu Street
Tel. 88129

Temple View
18 Poland Road
Tel. 649-444-9704

Upper Hutt
California Drive, Totara Park
Tel. 164-159

Wellington
140 Moxham Avenue
Hataitai
Tel. 861-227

SAMOA

Apia
No address listed
Tel. 160-685

TAHITI

Papeete
Avenue du Commandant Chesse
Tel 42-55-94

Pirae
B.P. 5682 Pirae
Tel. 43-94-68

Raiatea
No address listed
No phone listed

TONGA

Nuku´ Alofa
No address listed
Tel. 41-055

CENTRAL AND SOUTH AMERICA

ARGENTINA

Adrogué, Bs.As.
Quintana 447
No phone listed

Alta Gracia, Córdoba
Mescasni 236
No phone listed

Bahia Blanca, Bs.As.
Santa Fe and Pueyrredón
Tel. 44007

Banfield, Bs.As.
See Adrogué

Berazategui, Bs.As.
Calle 16 and Chile 138
No phone listed

Buenos Aires
See Adrogué, Berazategui, Capital
    Federal, San Fernando, San
    Miguel, and Vicente Lopez

Capital Federal
Calle Caracas 1289
No phone listed

Cordoba
See Alta Gracia

La Plata, Bs.As.
Calle 4 525
No phone listed

Litoral, Bs.As.
See San Miguel

Mar del Plata, Bs.As.
San Luis 3950
Tel. 728298

Mendoza, Mendoza
Calle Olascoaga 1015
Tel. 307252

Merlo, Bs.As.
See San Antonio de Padua

Quilmes, Bs.As.
See Berazategui

Resistencia, Chaco
Entre Rios and Linier
Tel. 0783-61471

Rosario, Rosario
Agrelo 1171
Tel. 387467

Salta, Salta
Mariano Boedo 56
Tel. 087-231741

San Antonio de Padua, Bs.As.
Curie 51
Tel. 22300

San Fernando, Bs.As.
1644 San Fernando
No phone listed

San Miguel, Bs.As.
Pringles 6207
No phone listed

Santa Fe, Santa Fe
Francia 3595
Tel. 20933

Santiago del Estero, Santiago del
  Estero
Rep. Del Líbano and Hernan-
  darias
No phone listed

Tucuman, Tucumán
Salta 450
Tel. 215071 208-785-5022

Vicente Lopez, Bs.As.
H. Irigoyen 1230
Tel. 701-4881

BRAZIL

Alegrete (RS)
Rua Waldemar Masson, 85
No phone listed

Araraquara (SP)
Rua Voluntários da Pátria, 1219
Tel. 0162-36-8242

Bauru (SP)
Rua Aparecida, 8-35
Tel. 0142-22-6194

Belo Horizonte (MG)
Rua Levindo Lopes, 214
Tel. 031-223-7883

Boa Viagem (PB)
Rua Sá e Souza, 580
Tel. 081-341-4333

Brasília (DF)
Av. W4 Norte, Q.712-A.ESP.
Tel. 061-274-8450

Campinas (SP)
See Jundiai

Campinas Castelo (SP)
Rua Albano de Almeida
Lima, 669
Tel. 0192-41-7545

Canoas (RS)
Rua Marques do Herval, 359
Tel. 0512-22-2712

Curitiba Iguaçu (PR)
Av. Iguaçu, 1460
Tel. 041-234-3972

Curitiba Sul (PR)
Praça Joseph Smith, 15
Tel. 041-244-4690

Florianopólis (SC)
Av. Delamar J. da Silva, 155
Sao José
Tel. 0474-23-4588

Fortaleza (Ceará)
Rua Neuma, 123
Vila Kennedy
Tel. 083-228-5107

João Pessoa (PB)
Rua João Amorim, 429
Tel. 083-222-1836

Joinville (SC)
Rua Max Colin, 426
Tel. 0474-33-5519

Jundiai (SP)
Rua Conde de Monsanto, 374
Tel. 0192-41-7545

Maceió (AL)
Av. Teresa Cristina, 287/307
Farol
Tel. 082-231-6633

Marília (SP)
See Bauru

Niteroi (RJ)
Rua Ribeiro Leite, 22
Tel. 021-717-4151

Novo Hamburgo (RS)
Rua Julio Aischinger, 80
Tel. 0512-93-7340

Petropólis (RJ)
Rua Souza Franco, 428
Tel. 0242-43-6460

Ponta Grossa (PR)
Av. Bonifacio Vilela, 460
Tel. 0422-23-4267

Porto Alegre (RS)
See Canoas

Recife
See Boa Viagem

Rio Claro (SP)
Av. Seis, 458
Tel. 0195-24-2844

Rio de Janeiro (RJ)
Rua Getulio, 49 - Meier
Tel. 021-269-2245

Santo André (SP)
Rua Catequese, 432
Tel. 011-449-3587

Santos (SP)
Av. Dr. Waldemar Leão, 305
Jabaquara
Tel. 0132-32-3962

São Paulo (SP)
Av. Prof. Francisco Morato, 2430
Tel. 011-814-3492

São Paulo Norte (SP)
Av. Nova Cantareira, 1146
Tucuruvi
Tel. 011-299-3893

Sorocaba (SP)
Rua Visconde do Rio Branco, 290
Tel. 0152-31-5408

Vitória (ES)
No address listed
Tel. 027-226-5268

CHILE

Republica
Corbea #2320
No phone listed

Santiago
Pedro de Valdivia #1423
Tel. 40083-741218

Viña Del Mar
3 Norte #1105
No phone listed

COLOMBIA

Bogotá
Carrera 38A #57B-26
No phone listed

COSTA RICA

Alajuela
Barrio la Tropicana
Tel. 41-29-47

EL SALVADOR

San Salvador
No address listed
No phone listed

GUATEMALA

Ciudad de Guatemala
3a Av. 14-16 Zona 1
Tel. 51-10-58

Quezaltenango
7a Calle 13-10 Zona 3
Tel. 061-2441

MEXICO

Celaya, Guanajuato
Fuente Ovejuna #673
No phone listed

Chihuahua, Chihuahua
Carbonel #1301
Tel. 14-13-11-92

Ciudad de Mexico
See San Juan de Aragón

Ciudad Madero, Tamaulipas
Ejército Nal. #75
Col. Loma del Gallo
No phone listed

Ciudad Obregon, Sonora
Veracruz and Yaqui
No phone listed

Ciudad Victoria, Tamaulipas
Calle 8 Anaya #1203
No phone listed

Colonia Guadalupe, Monclova
Mérida and Caracas #201
No phone listed

Colonia Juarez, Chihuahua
Av. Juarez 33
Tel 500-80

Hermosillo, Sonora
García Conde #303, Col. Pitic
No phone listed

Los Mochis, Sinaloa
Guillermo Prieto No. 442
Tel. 5-10-33

Matamoros, Tamaulipas
Marte #60
No phone listed

Mérida, Yucatán
Calle 70 #527, Int. 65 and 67
Tel. 1-05-36

Monclova
Mérida and Caracas #201
Col. Guadalupe
Tel. 32432

Monterrey, Nuevo León
Cerralvo #134, Guadalupe
No phone listed

Orizaba, Veracruz
Pte. 20 Esq. c/Nte. 3
No phone listed

Poza Rica, Veracruz
Hidalgo #501
No phone listed

Puebla, Puebla
25 Sur #907
Tel. 48-18-01

Reynosa, Tamaulipas
No address listed
No phone listed

San Juan de Aragón, Distrito
   Federal
Av. 510 #90
Tel. 5-551-7634

Tijuana, Baja California Norte
Matamoros Esq. c/Aldama #730
No phone listed

Valle Hermoso, Tamaulipas
No address listed
No phone listed

Veracruz, Veracruz
Altamirano #27
No phone listed

PERU

Lima
Av. Javier Prado Este, Esq.
c/ Av. de Los Ingenieros
Urb. Santa Patricia, 3ra Etapa,
   La Molima
Tel. 36-12-84 (Anexo 30)

Calle Manuel Gonzáles Olaechea
   #393
San Isidro
Tel. 41-13-36

Nicolás de Pierola #500
5ta etapa, Urb. Ingeniería
San Martín de Porras
No phone listed

VENEZUELA

Caracas
2da Avenida de Campo Alegre,
   No. 14
Urbanización Campo Alegre,
   Chacao
Apartado 62569, Caracas
No phone listed

## AFRICA AND ASIA

### SOUTH AFRICA

Capetown
See Mowbray

Durban
144 Silverton Road
Berea
Tel. 031-223024

Johannesburg
1 Hunter Street
Highlands
Tel. 011-618-1890

Mowbray
Main and Grove Road
Tel. 021-69-8718

### ZIMBABWE

Harare
67 Enterprise Road
Highland
No phone listed

### JAPAN

Tokyo
5-10-30 Minami-azabu
Minato-ku
Tel. 03-440-3244

### KOREA

Seoul
No address yet
No phone listed

### TAIWAN

Taipei
#5, Lane 183, Chin Hua Street
Tel. 02-3210690

#209 Fu Lin Road
Shih Lin District
No phone listed

Kaohshung
292 Shih Chung I Road
No phone listed

### GUAM

Barrigada
Route 40
No phone listed

### PHILIPPINES

Bacolad City
Mansilingan
No phone listed

Butuan
Montilla Boulevard
No phone listed

Cadiz City
Andrea Village Subdivision
No phone listed

Caloocan City
25 Greer Technical Street
University Hills
Tel. 362-28-85

Cebu City
47 Salinas Drive
Lahug
Tel. 96960

Cebu City
Raffinan Cpd., Sikatuna Street
Tel. 53450

Iligan
Camague, Lanao del Norte
No phone listed

Laguna
236 Banlic, Cabuyao
Tel. 566-2456

Legaspi
M. Marquez Street, Bgy15
Tel. 52-80

Makati
Buendia Avenue
Tel. 88-62-92

Marikina
St. Mary's Avenue
Provident Village
No phone listed

Paranaque
Pamplona Chapel
Dona Cecilia
No phone listed

Quezon City
Panay Avenue
Tel. 922-91-94

# Appendix
# 6

# Accredited Genealogists

This is a list of genealogists who have passed the accredited examinations given by the Family History Library of The Church of Jesus Christ of Latter-day Saints. They have demonstrated their knowledge and ability to do genealogical research. The genealogists are listed under their area of accreditation. You may contact a genealogist on this list to do research for you. The Family History Library does not recommend individual genealogists, nor is it responsible for the performance of the genealogist you hire.

### CONTACTING AN ACCREDITED GENEALOGIST

It is a good idea to contact more than one accredited genealogist to learn what their rates, time commitments, and reporting procedures are.

When writing an accredited genealogist, include:

1. A copy of your pedigree chart.
2. Details on those lines that need work.
3. Details on research that has already been done on those lines that need work. (Send copies of your documents only. Do not send originals.)
4. A self-addressed, stamped return envelope, or one with international reply coupon(s) if you do not live in the same country as the genealogist.

EASTERN STATES
*(Includes Delaware, New York, New Jersey, and Pennsylvania)*
Craig L. Albiston
RFD #1, Box 104A
Logan, Utah 84321
801-752-2903

James Black
2731 Loran Heights Drive
Salt Lake City, Utah 84109
801-467-8113

Carol H. Cannon
3718 Delia Circle
Salt Lake City, Utah 84109
801-278-5005

Bruce Despain
118 "R" Street
Salt Lake City, Utah 84103
801-533-9151

Kory L. Meyerink
2219 East 3020 South
Salt Lake City, Utah 84109
801-466-1888

Vonalee E. Murray
9903 Dewey Road
Waterford, Pennsylvania 16441

Jimmy B. Parker
935 North 325 West
Bountiful, Utah 84010
801-295-3898

James W. Petty
P.O. Box 893
Salt Lake City, Utah 84110
801-572-4049
(Also consultation and family histories)

Kip Sperry
P.O. Box 11381
Salt Lake City, Utah 84147
(Consultation only)

Dr. John F. Vallentine
425 WIDB
Brigham Young University
Provo, Utah 84602
801-489-6977 or 801-378-2278

Elaine Washburn
5275 S. Champagne Avenue
Salt Lake City, Utah 84118
801-969-3310

MIDWESTERN STATES
*(Includes Illinois, Minnesota, Indiana, Missouri, Iowa, Ohio, Michigan, and Wisconsin)*
Wilma Adkins
P.O. Box 11394
Salt Lake City, Utah 84147
807-278-4806

LaRene Gaunt
10122 Buttercup Drive
Sandy, Utah 84092
801-572-0169

Ruth Gomez de Schirmacher
1861 Downington
Salt Lake City, Utah 84108
801-485-0275

Harold M. Hegyessey, Jr.
1030 East Grove Drive
Pleasant Grove, Utah 84062
801-785-2925

DeLoris A. Hill
707 South 200 West #A
Brigham City, Utah 84302
801-723-3358

Donald L. Horner
P.O. Box 1147
Pocatello, Idaho 83204
208-237-3059

Rebecca Masters Huffman
620 Aspen Drive
Summit Park, Utah 84060
(Consultation only)

Kory L. Meyerink
2219 East 3020 South
Salt Lake City, Utah 84109
801-466-1888
(Also consultation and family
    histories)

Jimmy B. Parker
935 North 325 West
Bountiful, Utah 84010
801-295-3898

Mary Peters
329 North 200 West
Salt Lake City, Utah 84103
801-363-9585

Ellen Piehl
P.O. Box 9112
Green Bay, Wisconsin 54308
414-435-9294

Robert H. Raschig
1115 W. Petters Circle #806
Murray, Utah 84123
801-262-7708

Jay Roberts
1835-A Independence Boulevard
Salt Lake City, Utah 84116

Mary Schwartz
11946 Aneta Street
Culver City, California 90230
(Consultation only)

Kip Sperry
P.O. Box 11381
Salt Lake City, Utah 84147
(Consultation only)

Dr. John F. Vallentine
425 WIDB
Brigham Young University
Provo, Utah 84602
801-489-6977 or 801-378-2278

NEW ENGLAND STATES
*(Includes Connecticut, Maine, Mass-*
    *achusetts, New Hampshire, Rhode*
    *Island, and Vermont)*
Alice (Soule) Brower
188 South 100 East
Logan, Utah 84321
801-752-9035

Dr. Don K. Dalling
6285 South 725 East
Murray, Utah 84107
801-262-3630

Loren V. Fay
P.O. Box 2167
Albany, New York 12220
(Consultation only; also Quaker
    research)

Mrs. Lenna Fife
12681 West Reservation Road
Pocatello, Idaho 83202
208-237-0178

Elaine Justesen
8252 Derby Way
West Jordan, Utah 84084
801-561-2343

Mrs. M. Mary Litster
3256 South 675 West
Bountiful, Utah 84010
801-292-8918

Kory L. Meyerink
2219 East 3020 South
Salt Lake City, Utah 84109
801-466-1888

James W. Petty
P.O. Box 893
Salt Lake City, Utah 84110
801-572-4049
(Also consultation and family
    histories)

Robert N. Seaver
22420 Argus
Detroit, Michigan 48219
313-537-9366

Lee Smeal
P.O. Box 25743
Salt Lake City, Utah 84125
801-328-0787

Kip Sperry
P.O. Box 11381
Salt Lake City, Utah 84147
(Consultation only)

Clifford L. Stott
64 East 900 South
Orem, Utah 84058
801-226-0113

Eldon Walker
4218 Benview Drive
West Valley City, Utah 84120
801-969-3670

SOUTHERN STATES
*(Includes Mississippi, Alabama,*
    *North Carolina, Arkansas, Okla-*
    *homa, Florida, South Carolina,*
    *Georgia, Tennessee, Kentucky,*
    *Texas, Louisiana, Virginia, Mary-*
    *land, and West Virginia)*
Golden V. Adams, Jr.
780 West 1340 South
Provo, Utah 84601
801-375-3872
(Consultation only)

Wilma Adkins
P.O. Box 11394
Salt Lake City, Utah 84147
807-278-4806

Steven Blodgett
1161 Webster Drive
Sandy, Utah 84070
801-571-0308

Kathleen V. Cook
P.O. Box 1277
Escondido, California 92025

Jeanette B. Daniels
2142 Lambourne
Salt Lake City, Utah 84109
801-485-3446

Marie A. Davidson
P.O. Box 314
Burley, Idaho 83318

Marilyn Deputy
14343 Addison Street #223
Sherman Oaks, California 91423

Ernest B. Faulconer
1280 West Camelot Drive
Provo, Utah 84601
801-375-6997

JoAnn F. Hatch
127 North West Street
P.O. Box 368
Snowflake, Arizona 85937
602-536-7274

Virginia Hershey
5325 Wikiup Bridgeway
Santa Rosa, California 95404
707-544-3921

Wayne Morris
925 North 660 West
West Bountiful, Utah 84087
801-292-2692

Carolyn J. Nell
3126 Juniper Lane
Falls Church, Virginia 22044
703-536-6205

James W. Petty
P.O. Box 893
Salt Lake City, Utah 84110
801-572-4049
(Also consultation and family
    histories)

Lola L. Sorensen
5518 Revere Drive
Salt Lake City, Utah 84117
801-262-9675

Rodney E. Stucker
145 West 200 North #4
Salt Lake City, Utah 84103
801-355-6929
(Specializing in family history

research, development, and
publication)

David E. Swan
1750 East Sandpiper Circle #48
Sandy, Utah 84070
801-272-6152

William Thorndale
150 North 200 West #31
Salt Lake City, Utah 84103

Marcia J. Tripp
3250 North River Drive
Eden, Utah 84310
801-745-2453

Dr. T. R. Turk
P.O. Box 1314
Port Lavaca, Texas 77979

Dr. John F. Vallentine
425 WIDB
Brigham Young University
Provo, Utah 84602
801-489-6977 or 801-378-2278

Claudia Wagoner
8694 South 3720 West
West Jordan, Utah 84088
801-569-0956 (Consultation only)

Verl F. Weight
5343 Halsted Avenue
Carmichael, California 95608
916-483-5617

Lusarah B. Whittal
1136 South Granada Avenue
Alhambra, California 91801
213-284-3465

Lyle E. Wiggins
3719 West Hillsboro Circle
West Jordan, Utah 84084

BRITISH CANADA
William E. Arbuckle
1844 South 1700 East
Salt Lake City, Utah 84108
801-485-5162

Barry E. Kirk
P.O. Box 1161
Salt Lake City, Utah 84147
801-968-7629

FRENCH CANADA AND ACADIA
Mrs. Freida D. Child
3123 South Crestview Circle
Bountiful, Utah 84010
801-298-3056

Barry E. Kirk
P.O. Box 1161
Salt Lake City, Utah 84147
801-968-7629

Paul H. Munson
P.O. Box 462
Goodyear, Arizona 85338

LDS CHURCH RECORDS
*(Mormon lineages, 1830–present)*
Patricia I. Clark
6275 Barton Park Drive
West Jordan, Utah 84084
801-969-7015

David L. Grundvig
925 South 10th East
Salt Lake City, Utah 84105
801-369-1033
(Incl. photocopying, IGI batch

number tracing, and film
searching)

Patricia D. Heilpern
2505 Hartford Street
Salt Lake City, Utah 84106

Patricia N. Howard
3869 Chatterleigh Road
West Valley City, Utah 84120

Ella M. Soelberg
13336 144th Avenue
Grand Rapids, Michigan 49417
616-842-0742

AUSTRALIA
Derek F. Metcalfe
P.O. Box 2054
Salt Lake City, Utah 84110

Janet Reakes
160 Johnston Road
Bass Hill, NSW 2197
Australia
Tel. 011-61-02-727-0824

BELGIUM
John Van Weezep
111 South Hillside Gardens
 Drive
North Salt Lake, Utah 84054
801-292-5420
(For U.S./Canadian patrons only)

CHINA, TAIWAN, HONG KONG
Basil Yang
411 East Springhill Circle
North Salt Lake, Utah 84054
801-295-0276

CZECHOSLOVAKIA
Erik B. Christensen
123 Second Avenue #607
Salt Lake City, Utah 84103
801-359-4797

Gerald M. Haslam, Ph.D.
Brigham Young University
4500 HBLL
Provo, Utah 84602
801-378-4388

Linda K. Larson
1449 South 800 East
Orem, Utah 84058
801-255-5232

Olga K. Miller
855 Foothill Drive
Salt Lake City, Utah 84108
801-582-8013

Daniel M. Schlyter
2305 South 1360 West
Salt Lake City, Utah 84119

Ellinor Sorensen
1951 Woodside Drive
Salt Lake City, Utah 84117
801-277-4726

DENMARK
B. A. Borgesen
8358 South 1370 East
Sandy, Utah 84092
801-561-2321

Aurelia N. Clemons
3485 O'Bryant
Idaho Falls, Idaho 83401
208-523-6446

Charla Jensen
4635 South Loyola Street
Salt Lake City, Utah 84120
801-964-8403

Peer K. Kristensen
837 West 3600 South
Bountiful, Utah 84010
801-295-5744

Mrs. Inger P. Ludlow
558 "K" Street
Salt Lake City, Utah 84103
801-355-2098

Kim Melchior
15 South 300 East #5
Salt Lake City, Utah 84111
801-533-9321

Steven R. Parkes
5092 West Elma Street
West Valley City, Utah 84120
801-967-3867

William O. Pedersen
1223 East 3670 South
Salt Lake City, Utah 84106
801-265-0239

ENGLAND
Karen R. Baggs
860 North 900 West
Provo, Utah 84601
801-373-8116

Nel Lo Bassett
1055 East Hillcrest Drive
Springville, Utah 84663
801-489-6298

Trevor Ian Burborough
261 South 100 West
Kaysville, Utah 84037
801-544-3598

Gloria D. Chaston
31 East 2050 North
Provo, Utah 84601

Mrs. Freida D. Child
3213 South Crestview Circle
Bountiful, Utah 84010
801-298-3056

Arlene M. Denney
2666 North 200 East
North Ogden, Utah 84404
801-782-5770

Phill B. Dunn
6652 South 3270 West
West Jordan, Utah 84084
801-966-7731
(Especially London, nearby
    counties, and West Midlands)

Roger Flick
88 North 600 East
Orem, Utah 84057
(Consultation for family
    organizations only)

Jessie May Foster
700 Ben Lomond Avenue
Ogden, Utah 84403
801-479-7359

David E. Gardner
406 8th Avenue
Salt Lake City, Utah 84103
801-322-4313
(London only)

Mrs. Jessie Gardner
Sterling, Idaho 83279
208-397-4013

Vera Lucille Gifford
Oxford Manor, No. 301
124 East 1st Avenue
Salt Lake City, Utah 84103
801-942-4647

Jeanine H. Gray
1883 Edison
Salt Lake City, Utah 84115
801-756-2760

Lorna H. Hale
179 Dorchester Drive
Salt Lake City, Utah 84103
801-364-8174

Gillian O. Harrington
9714 South 285 East
Sandy, Utah 84070
801-571-1844

DeLoris A. Hill
707 South 200 West #A
Brigham City, Utah 84302
801-723-3358

Judith G. Ison
947 McClelland Street
Salt Lake City, Utah 84105
801-359-5762

Lance Jacob
3956 Zana Lane
Magna, Utah 84044
801-250-3466

Laureen R. Jaussi
284 East 400 South
Orem, Utah 84057

Edna B. Jones
4144 Morris Street
Salt Lake City, Utah 84119
801-968-9155

John M. Kitzmiller II
P.O. Box 2640
Salt Lake City, Utah 84110-2640
801-292-8731
(Especially military and
    heraldry)

Raymond W. Madsen
532 West 1550 North
Lehi, Utah 84043
801-768-9806

Derek F. Metcalfe
P.O. Box 2054
Salt Lake City, Utah 84110

Deon Morlock
1625 East Cherry Lane
Fruit Heights, Utah 84037
801-544-4980

Alan J. Phipps
424 "A" Street
Salt Lake City, Utah 84103
801-355-1662
(Pre-1800 preferred; migration
    problems; field trips to
    England)

Richard W. Price
2061 St. Mary's Drive
Salt Lake City, Utah 84108
801-531-0920
(Especially Norfolk; Colonial
    emigration; field trips to
    England)

Claude C. Richards
1059 Briar Avenue
Provo, Utah 84601
801-373-9523

Neal S. Southwick
Department of Religion
Ricks College
Rexburg, Idaho 83440

Doris R. Steen
507 East 1864 South
Orem, Utah 84057
801-225-4903

J. Grant Stevenson
230 West 1230 North
Provo, Utah 84601
801-374-9600

Wildra L. Welch
350 Star Crest Drive (1840 West)
Salt Lake City, Utah 84116
801-537-5333

FINLAND
Timothy Laitila Vincent
1841 West Morton Drive #F7
Salt Lake City, Utah 84116
801-355-9200

FRANCE
Barry E. Kirk
P.O. Box 11661
Salt Lake City, Utah 84147
801-968-7629

Mrs. Marlene Lee
9775 Clondike Court
Boise, Idaho 83709
208-322-1259

Yvette B. Longstaff
2627 East 10000 South
Sandy, Utah 84092
801-942-2125

GERMANY
Fredrick H. Barth
1628 Browning Avenue
Salt Lake City, Utah 84105
801-484-1639

Axel Borcherding
Wincklerstrasse 16
D-3052 Bad Nenndorf
Federal Republic of Germany
05723/2971

Inge Bork
5700 South China Clay Drive
Salt Lake City, Utah 84118
801-966-1519

Dr. John A. Dahl
560 East South Temple #604
Salt Lake City, Utah 84102
801-328-1391

Bruce Despain
118 "R" Street
Salt Lake City, Utah 84103
801-533-9151

Dr. Richard W. Dougherty
638 6th Avenue
Salt Lake City, Utah 84103
801-322-4610

Ruth E. Froelke
9717 Altamont Drive
Sandy, Utah 84092
801-942-4820

C. Russell Jensen, Ph.D.
3425 South Eastwood Drive
Salt Lake City, Utah 84109
(Specializing in Germanic-Latin)

Kory L. Meyerink
2219 East 3020 South
Salt Lake City, Utah 84109
801-466-1888
(Also consultation and family
    history)

Karl-Michael Sala
3668 Vistawest Drive
Box 1028
West Jordan, Utah 84084
801-569-8857
(Also East Germany field research)

Trudy Schenk
1230 Kensington Avenue
Salt Lake City, Utah 84105
801-467-8087
(Southern Germany only)

Marion Wolfert
2541 Campus Drive
Salt Lake City, Utah 84121
801-943-8891

Friedrich Wollmershauser
Stuttgarter Strasse 133
D-7261 Ostelsheim
Federal Republic of Germany
(Württemberg, if place of origin is
    known; Baden; Hohenzollern)

IRELAND
Judith E. Wight
276 West 1310 North
Orem, Utah 84057
801-224-2354

**ITALY**
Trafford R. Cole
Via Livenza 12
35010 Pionca di Vigonza (PD)
Italy

Dr. Giovanni Tata
P.O. Box 8414
Salt Lake City, Utah 84108

Mark T. Urban
P.O. Box 2151
Provo, Utah 84603
801-375-4883

**MEXICO**
Ruth Gomez de Schirmacher
1861 Downington Avenue
Salt Lake City, Utah 84108
801-485-0275

**NETHERLANDS**
Angeline (Eegelina Ylst) Hut
4316 Adonis Drive
Salt Lake City, Utah 84024
801-277-5921

Erica Nederhand
118 "N" Street
Salt Lake City, Utah 84103
801-359-3148

Christina Van Oostendorp
1377 East Jasmine
Sandy, Utah 84092
801-571-0360

John Van Weezep
111 S. Hillside Gardens Drive
North Salt Lake, Utah 84054
801-292-5420
(For U.S./Canadian patrons only)

**NEW ZEALAND**
Irene A. Davies
"Kawa"
c/o P.O. Okiwi
Great Barrier Island
New Zealand

Derek F. Metcalfe
P.O. Box 2054
Salt Lake City, Utah 84110

**NORWAY**
Karin Christensen
624 North 400 East
Kaysville, Utah 84037
801-544-1797

Elaine Helgeson Hasleton
322 East Barnard Street
Centerville, Utah 84014
801-292-5806

Ruth E. Maness
P.O. Box 293
Midvale, Utah 84047-0293
801-292-1045

Wade C. Starks
140 West 900 South
Orem, Utah 84057
801-225-6804

Mrs. Raeone C. Steuart
1987 Claremont Drive
Bountiful, Utah 84010
801-292-7168

**POLAND**
Zdenka Kucera
1270 Laird Avenue
Salt Lake City, Utah 84105
801-467-8611

POLYNESIA
Patricia I. Clark
6275 Barton Park Drive
West Jordan, Utah 84084
801-969-7015

Elwin W. Jensen
7347 South 2345 West
West Jordan, Utah 84084
801-566-4575

V. Foli P. Po'uha
786 Catherine Street
Salt Lake City, Utah 84116
801-539-8552

SCOTLAND
William L. Arbuckle
1844 South 1700 East
Salt Lake City, Utah 84108
801-485-5162

Jesse R. Gathercoal
686 East Cutler Avenue
Springville, Utah 84663
801-489-7716

Roberta L. Lindsay
30 North 600 East
Bountiful, Utah 84010
801-292-6095

Dean L. McLeod
1775 East Ellen Way
Sandy, Utah 84092
801-571-7884 or
801-378-4388 (out of state)

Ida H. Wagstaff
420 West 7th North
American Fork, Utah 84003

SOUTH AFRICA
Suzanne M. Scott
12599 South 1700 East
Draper, Utah 84020

SPAIN
George R. Ryskamp
4275 Edgewood Place
Riverside, California 92506

SWEDEN
Donald W. Christensen
553 Julep Drive
Murray, Utah 84107
801-262-2896

Dennis J. Hjelm
P.O. Box 646
Basalt, Idaho 83218
208-346-6188

Rolf B. Magnusson
3284 West 3595 South
Salt Lake City, Utah 84119
801-969-3079

Kim Melchior
15 South 300 East #5
Salt Lake City, Utah 84111
801-533-9321

Russell C. Robinson, Jr.
171 Clifton Avenue
West Hartford, Connecticut 06107
203-561-0851

Margareta Soderquist
2555 West 15000 South
Bluffdale, Utah 84065
801-254-1095

Nora J. Sutton
454 West 300 North
Logan, Utah 84321

SWITZERLAND
Dr. John A. Dahl
560 East South Temple #604
Salt Lake City, Utah 84102
801-328-1391

Ruby C. Waterlyn
165 Beryl Avenue
Salt Lake City, Utah 84115
801-487-2539

Mrs. Barbara Whiting
1334 Maple Lane
Provo, Utah 84604
801-377-4754

# Appendix
# 7

# Addresses for Driver's License Transcripts

## ALABAMA

Driver's License Division
Certification Section
P.O. Box 1471-H
Montgomery, Alabama 36192

## ALASKA

Department of Public Safety
Drivers' License Section
P.O. Box 20020-E
Juneau, Alaska 99802

## ARIZONA

Motor Vehicle Division
P.O. Box 2100-L
Phoenix, Arizona 85001

## ARKANSAS

Office of Drivers Services
Traffic Violation Reports
P.O. Box 1272-L
Little Rock, Arkansas 72203
(Write for required form)

## CALIFORNIA

Department of Motor Vehicles
P.O. Box 944231-0
Sacramento, California 94244

## COLORADO

Department of Revenue
Motor Vehicle Division
140 West Sixth Avenue
Denver, Colorado 80204

CONNECTICUT

Department of Motor Vehicles
Copy Record Section
60 State Street
Wethersfield, Connecticut 06109

DELAWARE

Motor Vehicles Department
P.O. Box 698-M
Dover, Delaware 19903

DISTRICT OF COLUMBIA

Department of Transportation
Bureau of Motor Vehicles
301 C Street NW
Washington, D.C. 20001

FLORIDA

Drivers' License Division
Department of Highway Safety
Kirkham Building
Tallahassee, Florida 32399
(Write for required form)

GEORGIA

Department of Public Safety
Drivers' Service Station
P.O. Box 1456-1
Atlanta, Georgia 30301

HAWAII

District of First Circuit Court
Violations Bureau
530 S. King Street
Honolulu, Hawaii 96813

IDAHO

Department of Law Enforcement
Motor Vehicle Division
P.O. Box 7129-T
Boise, Idaho 83707

ILLINOIS

Secretary of State
Drivers' Services Department
2701 South Dirksen Parkway
Springfield, Illinois 62723

INDIANA

Bureau of Motor Vehicles
Paid Mail Section
Room 416, State Office Building
Indianapolis, Indiana 46204

IOWA

Department of Transportation
Records Section
100 Euclid Avenue
Des Moines, Iowa 50306

KANSAS

Division of Vehicles
Driver Control Bureau
Docking Office Building
Topeka, Kansas 66626

KENTUCKY

Division of Driver Licensing
State Office Building
Frankfort, Kentucky 40622

LOUISIANA

Department of Public Safety
Drivers' License Division
O.D.R. Section, Box 64886-C
Baton Rouge, Louisiana 70896

**MAINE**

Motor Vehicle Division
State House, Room 29
Augusta, Maine 04333

**MARYLAND**

Motor Vehicle Administration
6601 Ritchie Highway
Glen Burnie, Maryland 21062

**MASSACHUSETTS**

Registry of Motor Vehicles
Court Records Section
100 Nashua Street
Boston, Massachusetts 02114

**MICHIGAN**

Department of State
Bureau of Driver Services
Commercial Look-up Unit
7064 Crowner Drive
Lansing, Michigan 48918

**MINNESOTA**

Department of Public Safety
Drivers' License Office
Room 108, State Highway
  Building
St. Paul, Minnesota 55155
(Write for required form)

**MISSISSIPPI**

Highway Safety Patrol
Drivers' License Issuance Board
P.O. Box 958-H
Jackson, Mississippi 39205

**MISSOURI**

Bureau of Drivers' Licenses
Department of Revenue
P.O. Box 200-E
Jefferson City, Missouri 65105

**MONTANA**

Highway Patrol
303 Roberts
Helena, Montana 59620

**NEBRASKA**

Department of Motor Vehicles
Drivers' Records Section
301 Centennial Mall
Lincoln, Nebraska 68509

**NEVADA**

CMB of Nevada
555 Wright Way
Carson City, Nevada 89711

**NEW HAMPSHIRE**

Division of Motor Vehicles
Driver Record Unit
10 Hazen Drive
Concord, New Hampshire 03305

**NEW JERSEY**

Division of Motor Vehicles
Bureau of Security Responsibility
25 South Montgomery Street
Trenton, New Jersey 08666

**NEW MEXICO**

Transportation Department
Driver Service Bureau
P.O. Box 1028-L
Santa Fe, New Mexico 87502

NEW YORK

Department of Motor Vehicles
Public Service Bureau
Empire State Plaza
Albany, New York 12228

NORTH CAROLINA

Traffic Records Section
1100 New Bern Avenue
Raleigh, North Carolina 27697

NORTH DAKOTA

Drivers' License Division
Capitol Grounds
Bismarck, North Dakota 58505

OHIO

Bureau of Motor Vehicles
P.O. Box 7167-L
Columbus, Ohio 43266

OKLAHOMA

Drivers' Record Service
Department of Public Safety
P.O. Box 11415-F
Oklahoma City, Oklahoma 73136

OREGON

Motor Vehicles Division
1905 Lona Avenue
Salem, Oregon 97314

PENNSYLVANIA

Department of Transportation
Operator Information Section
P.O. Box 8695-L
Harrisburg, Pennsylvania 17105

RHODE ISLAND

Registry of Motor Vehicles
345 Harris Avenue
Providence, Rhode Island 02903

SOUTH CAROLINA

Department of Highways and
    Public Transportation
Department of Highways
    Building
Driver Record Check Section
P.O. Box 1498-R
Columbia, South Carolina 29216

SOUTH DAKOTA

Department of Public Safety
Driver Improvement Program
118 West Capitol
Pierre, South Dakota 57501
(Write for required form)

TENNESSEE

Department of Safety
P.O. Box 945-E
Nashville, Tennessee 37202

TEXAS

Department of Public Safety
License Issuance and Drivers'
    Records
P.O. Box 4087-S
Austin, Texas 78773
(Write for required form)

UTAH

Drivers' License Division
1095 Motor Avenue
Salt Lake City, Utah 84116

**VERMONT**

Agency of Transportation
Department of Motor Vehicles
Montpelier, Vermont 05603

**VIRGINIA**

Division of Motor Vehicles
Driver Licensing & Information
2300 W. Broad Street
Richmond, Virginia 23269

**WASHINGTON**

Division of Licensing
211 12th Avenue, S.E.
Olympia, Washington 98504

**WEST VIRGINIA**

Driver Improvement Division
Department of Motor Vehicles
1800 Washington Street, East
Charleston, West Virginia 25317

**WISCONSIN**

Department of Transportation
Driver Record File
P.O. Box 7918
Madison, Wisconsin 53707

**WYOMING**

Department of Revenue
2200 Carey Avenue
Cheyenne, Wyoming 82002

# Appendix
## 8

# Addresses for Birth, Death, Marriage, and Divorce Certificates

ALABAMA

*Birth and Death*

Center For Health
State Department of Public Health
Montgomery, Alabama 36130-1701

*Marriage*

Same as Birth and Death

*Divorce*

Same as Birth and Death

ALASKA

*Birth and Death*

Department of Health and Social
    Services
Bureau of Vital Statistics
P.O. Box H-02G
Juneau, Alaska 99811-0675

*Marriage*

Same as Birth and Death

*Divorce*

Same as Birth and Death

**AMERICAN SAMOA**

*Birth and Death*

Registrar of Vital Statistics
Vital Statistics Section
Government of American Samoa
Pago Pago, American Samoa 96799

*Marriage*

Same as Birth and Death

*Divorce*

High Court of American Samoa
Tutuila, American Samoa 96799

**ARIZONA**

*Birth (long form)*
  *Birth (short form)*
  *Birth registration*
    *card*

Vital Records Section
Arizona Department of Health
  Services
P.O. Box 3887

*Death*

Phoenix, Arizona 85030

*Marriage*

Same as Birth and Death

*Divorce*

Same as Birth and Death

**ARKANSAS**

*Birth and Death*

Division of Vital Records
Arkansas Department of Health
4815 West Markham Street
Little Rock, Arkansas 72201

*Marriage*

Same as Birth and Death

*Divorce*

Same as Birth and Death

**CALIFORNIA**

*Birth and Death*

Vital Statistics Section
Department of Health Services
410 N Street
Sacramento, California 95814

*Marriage*

Same as Birth and Death

*Divorce*

Same as Birth and Death

**CANAL ZONE**

*Birth and Death*               Panama Canal Commission
                               Vital Statistics Clerk
                               APO Miami, Florida 34011

*Marriage*                     Same as Birth and Death

*Divorce*                      Same as Birth and Death

**COLORADO**

*Birth and Death*               Vital Records Section
                               Colorado Department of Health
                               4210 East 11th Avenue
                               Denver, Colorado 80220

*Marriage*                     Same as Birth and Death

*Divorce*                      Same as Birth and Death

**CONNECTICUT**

*Birth and Death*               Department of Health Services
                               Vital Records Section
                               Division of Health Statistics
                               State Department of Health Services
                               150 Washington Street
                               Hartford, Connecticut 06106

*Marriage*                     Same as Birth and Death

*Divorce*                      Same as Birth and Death

**DELAWARE**

*Birth and Death*               Bureau of Vital Statistics
                               Division of Public Health
                               P.O. Box 637
                               Dover, Delaware 19903

*Marriage*                     Same as Birth and Death

*Divorce*                      Same as Birth and Death

**DISTRICT OF COLUMBIA**

*Birth and Death*                    Vital Records Branch
                                     4251 Street NW, Room 3009
                                     Washington, D.C. 20001

*Marriage*                           Marriage Bureau
                                     515 5th Street NW
                                     Washington, D.C. 20001

*Divorce*                            Clerk, Superior Court for the
                                         District of Columbia, Family
                                         Division
                                     500 Indiana Avenue NW
                                     Washington, D.C. 20001


**FLORIDA**

*Birth and Death*                    Department of Health and
                                         Rehabilitation Services
                                     Office of Vital Statistics
                                     1217 Pearl Street
                                     Jacksonville, Florida 32202

*Marriage*                           Same as Birth and Death

*Divorce*                            Same as Birth and Death


**GEORGIA**

*Birth and Death*                    Georgia Department of Human
                                         Resources
                                     Vital Records Unit
                                     Room 217-H
                                     47 Trinity Avenue, SW
                                     Atlanta, Georgia 30334

*Marriage*                           Same as Birth and Death

*Divorce*                            Same as Birth and Death

GUAM

*Birth and Death*

Office of Vital Statistics
Department of Public Health and
  Social Services
Government of Guam
P.O. Box 2816
Agana, Guam, Marianas Islands
  96910

*Marriage*

Same as Birth and Death

*Divorce*

Same as Birth and Death

HAWAII

*Birth and Death*

Research and Statistics Office
State Department of Health
P.O. Box 3378
Honolulu, Hawaii 96801

*Marriage*

Same as Birth and Death

*Divorce*

Same as Birth and Death

IDAHO

*Birth and Death*

Bureau of Vital Statistics, Stan-
  dards, and Local Health Services
State Department of Health and
  Welfare
450 W. State Street
Boise, Idaho 83720-9990

*Marriage*

Same as Birth and Death

*Divorce*

Same as Birth and Death

ILLINOIS

*Birth and Death*

Office of Vital Records
State Department of Public Health
650 West Jefferson Street
Springfield, Illinois 62702-5079

| | |
|---|---|
| *Marriage* | Same as Birth and Death |
| *Divorce* | Same as Birth and Death |

**INDIANA**

*Birth and Death*  Division of Vital Records
State Board of Health
1330 West Michigan Street
P.O. Box 1964
Indianapolis, Indiana 46206-1964

*Marriage*  Same as Birth and Death

*Divorce*  Same as Birth and Death

**IOWA**

*Birth and Death*  Iowa State Department of Health
Vital Records Section
Lucas State Office Building
Des Moines, Iowa 50319

*Marriage*  Same as Birth and Death

*Divorce*  Same as Birth and Death

**KANSAS**

*Birth and Death*  Bureau of Vital Statistics
900 Southwest Jackson
Topeka, Kansas 66612-1290

*Marriage*  Same as Birth and Death

*Divorce*  Same as Birth and Death

**KENTUCKY**

*Birth and Death*  Office of Vital Statistics
Department for Human Resources
275 East Main Street
Frankfort, Kentucky 40621

*Marriage*  Same as Birth and Death

*Divorce*  Same as Birth and Death

LOUISIANA

*Birth (long form)*                 Division of Vital Records
*Birth (short form)*                Office of Health Services and
*Death*                                       Environmental Quality
                                            325-Loyola Avenue
                                            New Orleans, Louisiana 70112


*Marriage*
  *Orleans Parish*              Same as Birth and Death
  *Other Parishes*             Same as Birth and Death

*Divorce*                             Same as Birth and Death

MAINE

*Birth and Death*               Office of Vital Records
                                            Human Services Building
                                            Station II
                                            State House
                                            Augusta, Maine 04333


*Marriage*                          Same as Birth and Death

*Divorce*                            Same as Birth and Death

MARYLAND

*Birth and Death*              Division of Vital Records
                                           State Department of Health and
                                               Mental Hygiene
                                           Metro Building
                                           4201 Patterson Avenue
                                           P.O. Box 68760
                                           Baltimore, Maryland 21215-0020


*Marriage*                         Same as Birth and Death

*Divorce*                           Same as Birth and Death

MASSACHUSETTS

*Birth and Death*             Registry of Vital Records and
                                              Statistics
                                          150 Tremont Street, Room B-3
                                          Boston, Massachusetts 02111

| | |
|---|---|
| *Marriage* | Same as Birth and Death |
| *Divorce* | Same as Birth and Death |

**MICHIGAN**

*Birth and Death*

Office of the State Registrar
Center of Health Statistics
Michigan Department of Public
   Health
3500 North Logan Street
Lansing, Michigan 48909

*Marriage*

Same as Birth and Death

*Divorce*

Same as Birth and Death

**MINNESOTA**

*Birth and Death*

Minnesota Department of Health
Section of Vital Statistics
717 Delaware Street SE
P.O. Box 9441
Minneapolis, Minnesota 55440

*Marriage*

Same as Birth and Death

*Divorce*

Same as Birth and Death

**MISSISSIPPI**

*Birth and Death*

Vital Records
State Board of Health
North State Street
Jackson, Mississippi 39216

*Marriage*

Same as Birth and Death

*Divorce*

Same as Birth and Death

**MISSOURI**

*Birth and Death*

Division of Health
Bureau of Vital Records
1730 E. Elm
Jefferson City, Missouri 65102

| | |
|---|---|
| *Marriage* | Same as Birth and Death |
| *Divorce* | Same as Birth and Death |

MONTANA

| | |
|---|---|
| *Birth and Death* | Bureau of Records and Statistics<br>State Department of Health and<br>    Environmental Sciences<br>Helena, Montana 59620 |
| *Marriage* | Same as Birth and Death |
| *Divorce* | Same as Birth and Death |

NEBRASKA

| | |
|---|---|
| *Birth and Death* | Bureau of Vital Statistics<br>State Department of Health<br>301 Centennial Mall South<br>P.O. Box 95007<br>Lincoln, Nebraska 68509-5007 |
| *Marriage* | Same as Birth and Death |
| *Divorce* | Same as Birth and Death |

NEVADA

| | |
|---|---|
| *Birth and Death* | Division of Health and Vital<br>    Statistics<br>Capitol Complex<br>Carson City, Nevada 89710 |
| *Marriage* | Same as Birth and Death |
| *Divorce* | Same as Birth and Death |

NEW HAMPSHIRE

| | |
|---|---|
| *Birth and Death* | Bureau of Vital Records<br>Health and Welfare Building<br>Hazen Drive<br>Concord, New Hampshire 03301 |

| | |
|---|---|
| *Marriage* | Same as Birth and Death |
| *Divorce* | Same as Birth and Death |

NEW JERSEY

| | |
|---|---|
| *Birth and Death* | State Department of Health<br>Bureau of Vital Statistics<br>CN 370<br>Trenton, New Jersey 08625<br>or<br>Archives and History Bureau<br>State Library Division<br>State Department of Education<br>Trenton, New Jersey 08625 |
| *Marriage* | Same as Birth and Death |
| *Divorce* | Superior Court, Chancery Division<br>State House Annex, Room 320<br>CN 971<br>Trenton, New Jersey 08625 |

NEW MEXICO

| | |
|---|---|
| *Birth and Death* | Vital Statistics Bureau<br>New Mexico Health Services<br>Division<br>1190 St. Francis Drive<br>Santa Fe, New Mexico 87503 |
| *Marriage* | Same as Birth and Death |
| *Divorce* | Same as Birth and Death |

NEW YORK (EXCEPT NEW YORK CITY)

| | |
|---|---|
| *Birth and Death* | Bureau of Vital Records<br>State Department of Health<br>Empire State Plaza<br>Tower Building<br>Albany, New York 12237-0023 |
| *Marriage* | Same as Birth and Death |
| *Divorce* | Same as Birth and Death |

## NEW YORK CITY

| | |
|---|---|
| *Birth and Death* | Bureau of Vital Records<br>Department of Health of New York City<br>125 Worth Street<br>New York, New York 10013 |
| *Marriage*<br>*Bronx* | Marriage License Bureau<br>1780 Grand Concourse<br>Bronx, New York 10457 |
| *Brooklyn* | Marriage License Bureau<br>Municipal Building<br>Brooklyn Borough Hall<br>Brooklyn, New York 11201 |
| *Manhattan* | Marriage License Bureau<br>No. 1 Center Street<br>Municipal Building<br>New York, New York 10007 |
| *Queens* | Marriage License Bureau<br>Queens Borough Hall<br>120-55 Queens Boulevard<br>Kew Gardens, New York 11424 |
| *Staten Island (no longer called Richmond)* | Marriage License Bureau<br>Staten Island Borough Hall<br>St. George<br>Staten Island, New York 10301 |
| *Divorce* | Same as Marriage for all five boroughs |

## NORTH CAROLINA

| | |
|---|---|
| *Birth and Death* | Department of Human Resources<br>Division of Health Services<br>Vital Records Branch<br>225 North McDowell Street<br>P.O. Box 27687<br>Raleigh, North Carolina 27611-7687 |

*Marriage*                          Same as Birth and Death

*Divorce*                           Same as Birth and Death

NORTH DAKOTA

*Birth and Death*                   Division of Vital Records
                                    State Department of Health
                                    600 East Blvd. Avenue
                                    Bismarck, North Dakota 58505

*Marriage*                          Same as Birth and Death

*Divorce*                           Same as Birth and Death

OHIO

*Birth and Death*                   Division of Vital Statistics
                                    Ohio Department of Health
                                    G-20 Ohio Department Building
                                    65 South Front Street
                                    Columbus, Ohio 43266-0333

*Marriage*                          Same as Birth and Death

*Divorce*                           Same as Birth and Death

OKLAHOMA

*Birth and Death*                   Vital Records Section
                                    State Department of Health
                                    Northeast 10th Street and Stonewall
                                    P.O. Box 53551
                                    Oklahoma City, Oklahoma 73152

*Marriage*                          Same as Birth and Death

*Divorce*                           Same as Birth and Death

OREGON

*Birth and Death*                   Oregon State Health Division
                                    Vital Statistics Section
                                    P.O. Box 116
                                    Portland, Oregon 97207

| | |
|---|---|
| *Marriage* | Same as Birth and Death |
| *Divorce* | Same as Birth and Death |

**PENNSYLVANIA**

| | |
|---|---|
| *Birth and Death* | Division of Vital Statistics<br>State Department of Health<br>Central Building<br>101 South Mercer Street<br>P.O. Box 1528<br>New Castle, Pennsylvania 16103 |
| *Marriage* | Same as Birth and Death |
| *Divorce* | Same as Birth and Death |

**PUERTO RICO**

| | |
|---|---|
| *Birth and Death* | Division of Demographic Registry<br>and Vital Statistics<br>Department of Health<br>P.O. Box 11854<br>San Juan, Puerto Rico 00908 |
| *Marriage* | Same as Birth and Death |
| *Divorce* | Same as Birth and Death |

**RHODE ISLAND**

| | |
|---|---|
| *Birth and Death* | Division of Vital Statistics<br>State Department of Health<br>Room 101, Cannon Building<br>75 Davis Street<br>Providence, Rhode Island 02908 |
| *Marriage* | Same as Birth and Death |
| *Divorce* | Clerk of Family Court<br>1 Dorrance Plaza<br>Providence, Rhode Island 02903 |

**SOUTH CAROLINA**

*Birth and Death*

Office of Vital Records and Public
  Health Statistics
South Carolina Department of
  Health and Environmental Control
2600 Bull Street
Columbia, South Carolina 29201

*Marriage*

Same as Birth and Death

*Divorce*

Same as Birth and Death

**SOUTH DAKOTA**

*Birth and Death*

State Department of Health
Health Statistics Program
523 E. Capitol
Pierre, South Dakota 57501

*Marriage*

Same as Birth and Death

*Divorce*

Same as Birth and Death

**TENNESSEE**

*Birth and Death*

Tennessee Vital Records
Department of Health and
  Environment
Cordell Hull Building
Nashville, Tennessee 37219-5402

*Marriage*

Same as Birth and Death

*Divorce*

Same as Birth and Death

**TEXAS**

*Birth and Death*

Bureau of Vital Statistics
Texas Department of Health
1100 West 49th Street
Austin, Texas 78756-3191

*Marriage*

Same as Birth and Death

*Divorce*

Same as Birth and Death

**TRUST TERRITORY OF THE PACIFIC ISLANDS**

*Birth and Death*
  *Commonwealth of*             Commonwealth Courts
  *Northern Mariana*             Commonwealth Governments
  *Islands*                      Saipan, CM 96950

  *Republic of the*              Chief Clerk of Supreme Courts
    *Marshall Islands*           Republic of the Marshall Islands
                           Majuro, Marshall Islands 96960

  *Republic of Palau*          Chief Clerk of Supreme Courts
                           Republic of Palau
                           Koror, Palau, W.C.I. 96940

  *Federated States of*        Clerk of Courts
    *Micronesia*                State of Truk, FSM
                           Moen, Truk, E.C.I. 96942
                           or
                           Clerk of Courts
                           State of Ponape, FSM
                           Kolonia, Ponape, E.C.I. 96941
                           or
                           Clerk of Courts
                           State of Kosrae, FSM
                           Lele, Kosrae, E.C.I. 96944
                           or
                           Clerk of Courts
                           State of Yap, FSM
                           Colonia, Yap, W.C.I. 96943

*Marriage*                    Same as Birth and Death for all
                             territories

*Divorce*                      Same as Birth and Death for all
                             territories

**UTAH**

*Birth and Death*           Bureau of Health Statistics
                           Utah Department of Health
                           150 West North Temple
                           P.O. Box 16700
                           Salt Lake City, Utah 84116-0700

| | |
|---|---|
| *Marriage* | Same as Birth and Death |
| *Divorce* | Same as Birth and Death |

**VERMONT**

| | |
|---|---|
| *Birth and Death* | Vermont Department of Health<br>Vital Records Section<br>Box 70<br>60 Main Street<br>Burlington, Vermont 05402 |
| *Marriage* | Same as Birth and Death |
| *Divorce* | Same as Birth and Death |

**VIRGINIA**

| | |
|---|---|
| *Birth and Death* | Bureau of Vital Records<br>State Department of Health<br>P.O. Box 1000<br>Richmond, Virginia 23208-1000 |
| *Marriage* | Same as Birth and Death |
| *Divorce* | Same as Birth and Death |

**U.S. VIRGIN ISLANDS**

*Birth and Death*

| | |
|---|---|
| St. Croix | Registrar of Vital Statistics<br>Charles Harwood Memorial<br>Hospital<br>St. Croix, Virgin Islands 00820 |
| St. Thomas and<br>St. John | Registrar of Vital Statistics<br>Charlotte Amalie<br>St. Thomas, Virgin Islands 00802 |
| *Marriage* | Bureau of Vital Records and<br>Statistical Services<br>Virgin Islands Department of<br>Health<br>Charlotte Amalie<br>St. Thomas, Virgin Islands 00801 |

| | |
|---|---|
| *St. Croix* | Chief Deputy Clerk<br>Territorial Court of the Virgin<br>    Islands<br>P.O. Box 929<br>Christiansted<br>St. Croix, Virgin Islands 00820 |
| *St. Thomas and<br>    St. John* | Clerk of the Territorial Court of the<br>    Virgin Islands<br>P.O. Box 70<br>Charlotte Amalie<br>St. Thomas, Virgin Islands 00801 |
| *Divorce* | Same as Marriage |
| *St. Croix* | Same as Marriage |
| *St. Thomas and<br>    St. John* | Same as Marriage |

WASHINGTON

| | |
|---|---|
| *Birth and Death* | Vital Records<br>1112 South Quince<br>P.O. Box 9709, ET-11<br>Olympia, Washington 98504-9709 |
| *Marriage* | Same as Birth and Death |
| *Divorce* | Same as Birth and Death |

WEST VIRGINIA

| | |
|---|---|
| *Birth and Death* | Division of Vital Statistics<br>State Department of Health<br>State Office Building No. 3<br>Charleston, West Virginia 25305 |
| *Marriage* | Same as Birth and Death |
| *Divorce* | Same as Birth and Death |

**WISCONSIN**

*Birth and Death*

Bureau of Health Statistics
1 West Wilson Street
P.O. Box 309
Madison, Wisconsin 53701

**WYOMING**

*Birth and Death*

Vital Records
Hathaway Building
Cheyenne, Wyoming 82002

*Marriage*

Same as Birth and Death

*Divorce*

Same as Birth and Death

# Appendix
## 9

# STATE ADOPTION RECORDS

**RECORDS OPEN**

Alabama
Kansas
All others are closed

**COURT JURISDICTIONS**

See State-by-State Listings

**ORIGINAL BIRTH CERTIFICATES**

*Available to Adoptees*

Alabama
Kansas
Maine
Ohio

**OFFICIAL AGES OF MAJORITY**

|  | F | M |
|---|---|---|
| Alabama (both sexes) 18 | | |
| Colorado | 18 | 21 |
| Minnesota (both sexes) 21 | | |
| Mississippi | 18 | 21 |
| N. Dakota | 18 | 21 |
| Wyoming (both sexes) 19 | | |
| All other states 18 | | |

**ALABAMA**

Information Department
State Capitol
Montgomery, Alabama 36130

Bureau of Children's Services
Administration Building
604 North Union Street
Montgomery, Alabama 36130

Court of Jurisdiction
Probate

Birth and Death Records
Bureau of Vital Statistics
Department of Public Health
Montgomery, Alabama 13630

Marriage Records
Same as Birth and Death

Divorce Records
Same as Birth and Death

Archives/Records
Department of Archives
624 Washington Avenue
Montgomery, Alabama 36104


ALASKA

Information Department
State Capitol
120 4th Street
Juneau, Alaska 99811

State Agency
Department of Health Services
Pouch H-05
Juneau, Alaska 99811

Department of Social Services
Pouch H-05
Juneau, Alaska 98811

Court of Jurisdiction
Superior Court

Birth and Death Records
Bureau of Vital Statistics
Department of Health Services
Pouch H-05
Juneau, Alaska 99801

Marriage Records
Same as Birth and Death

Archives/Records
State Library
State Office Building
Pouch G
Juneau, Alaska 99801

ARIZONA

Information Department
State Capitol
1700 West Washington Street
Phoenix, Arizona 85007

State Agency
Department of Economic Security
Administration for Children
P.O. Box 6123
1400 West Washington Street
Phoenix, Arizona 85005

Court of Jurisdiction
Superior Court

Birth and Death Records
Division of Vital Records
Department of Health
P.O. Box 3887
Phoenix, Arizona 85030

Marriage Records
Clerk of Superior Court
(County issuing license)

Divorce Records
Clerk of Superior Court
(County where granted)

Archives/Records
State Library
State Capitol
1700 West Washington Street
Phoenix, Arizona 85007

ARKANSAS

Information Department
State Capitol
5th and Woodland
Little Rock, Arkansas 72201

State Agency
Department of Human Services
Adoption Services
P.O. Box 1437
Little Rock, Arkansas 72203

Court of Jurisdiction
Probate Court

Birth and Death Records
Division of Vital Records
4815 West Markham Street
Little Rock, Arkansas 72201

Marriage Records
Same as Birth and Death

Divorce Records
Same as Birth and Death

Archives/Records
State Library
One Capitol Mall
Little Rock, Arkansas 72201


CALIFORNIA

State Information Department
10th and L Streets
Sacramento, California 95814

State Agency
Department of Social Services
Adoption Branch
744 P Street
Sacramento, California 95814

Court of Jurisdiction
Superior Court

Birth and Death Records
Vital Statistics
410 N Street
Sacramento, California 95814

Marriage Records
Same as Birth and Death

Divorce Records
Same as Birth and Death

Archives/Records
State Library
914 Capitol Mall
P.O. Box 2037
Sacramento, California 95814


COLORADO

Information Department
State Capitol
200 East Colfax Street
Denver, Colorado 80203

State Agency
Department of Social Services
1575 Sherman Street
Denver, Colorado 80203

Court of Jurisdiction
District Court

Birth and Death Records
Statistics Section
4210 East 11th Avenue
Denver, Colorado 80220

Marriage & Divorce Records
Same as Birth and Death

Archives/Records
Colorado Heritage Center
1300 Broadway
Denver, Colorado 80203


CONNECTICUT

Information Department
210 Capitol Avenue
Hartford, Connecticut 06115

State Agency
Adoption Resource Exchange
170 Sigourney Street
Hartford, Connecticut 06105

Court of Jurisdiction
Probate Court

Birth and Death Records
Public Health Statistics Section
79 Elm Street
Hartford, Connecticut 06115

Marriage and Divorce Records
Same as Birth and Death

Archives/Records
State Library
Capitol Avenue
Hartford, Connecticut 06115


DELAWARE

Information Department
State Capitol
Dover, Delaware 19899

State Agency
Department of Health and Social
  Services
P.O. Box 309
Wilmington, Delaware 19899

Court of Jurisdiction
Superior Court

Birth and Death Records
Bureau of Vital Statistics
State Health Building
Dover, Delaware 19901

Marriage and Divorce Records
Same as Birth and Death

Archives/Records
Historical and Cultural Affairs
Department of State
Dover, Delaware 19901


DISTRICT OF COLUMBIA

Information Department
District Building
14th and E Streets
Washington, D.C. 20004

District Agency
Bureau of Family Services
122 C Street NW
Washington, D.C. 20001

Court of Jurisdiction
Domestic Relations Branch of
  Court of General Sessions

Birth and Death Records
Department of Human
  Resources
Vital Records Section
615 Pennsylvania Avenue
Washington, D.C. 20004

Marriage Records
Marriage Bureau
515 5th Street NW
Washington, D.C. 20001

Divorce Records
Clerk-Superior Court for D.C.
451 Indiana Avenue
Washington, D.C. 20001

Archives/Records
Library of Congress Annex
Washington, D.C. 20540

FLORIDA

Information Department
The Capitol
Tallahassee, Florida 32304

State Agency
Department of Health
Children's Services
1317 Winewood Boulevard
Tallahassee, Florida 32301

Court of Jurisdiction
Circuit Court

Birth and Death Records
Bureau of Vital Statistics
P.O. Box 210
Jacksonville, Florida 32231

Marriage and Divorce Records
Same as Birth and Death

Archives/Records
State Library
R. A. Gray Building
Tallahassee, Florida 32301

GEORGIA

Information Department
State Capitol
Capitol Square SW
Atlanta, Georgia 30334

State Agency
Division of Social Services
618 Ponce de Leon Avenue NE
Atlanta, Georgia 30308

Court of Jurisdiction
Superior Court

Birth and Death Records
Vital Records Unit
Department of Human Resources
Room 217-H
47 Trinity Avenue SW
Atlanta, Georgia 30334

Marriage and Divorce Records
Same as Birth and Death

Archives/Records
State Library
Capitol Hill Station
Atlanta, Georgia 30334

HAWAII

Information Department
State Capitol
415 South Beretania Street
Honolulu, Hawaii 96813

State Agency
Department of Social Services
P.O. Box 339
Honolulu, Hawaii 96809

Court of Jurisdiction
Family Court

Birth and Death Records
Research and Statistics
P.O. Box 3378
Honolulu, Hawaii 96801

Marriage and Divorce Records
Same as Birth and Death

Archives/Records
State Archives
Iolani Palace Grounds
Honolulu, Hawaii 96813

### IDAHO

Information Department
Statehouse
700 West Jefferson Street
Boise, Idaho 83720

State Agency
Department of Health and
  Welfare
Statehouse
Boise, Idaho 83720

Court of Jurisdiction
Magistrate Court

Birth and Death Records
Bureau of Vital Statistics
Statehouse
Boise, Idaho 83720

Marriage and Divorce Records
Same as Birth and Death

Archives/Records
Library and Archives Building
325 West State
Boise, Idaho 83720

### ILLINOIS

Information Department
State Capitol
Springfield, Illinois 62706

State Agency
Children and Family Services
One North Old State Capitol
  Plaza
Springfield, Illinois 62706

Court of Jurisdiction
County Circuit Court

Birth and Death Records
Vital Records Office
535 West Jefferson Street
Springfield, Illinois 62761

Marriage and Divorce Records
Same as Birth and Death

Archives/Records
State Archives Building
Springfield, Illinois 62706

### INDIANA

Information Department
200 West Washington Street
Indianapolis, Indiana 46204

State Agency
Department of Public Welfare
141 South Meridian Street
Indianapolis, Indiana 46225

Court of Jurisdiction
County Court

Birth and Death Records
Vital Records
1330 West Michigan Street
Indianapolis, Indiana 46206

Marriage and Divorce Records
Same as Birth and Death

Archives/Records
State Library
140 North Senate Avenue
Indianapolis, Indiana 46204

IOWA

Information Department
Capitol Building
1007 East Grand Avenue
Des Moines, Iowa 50319

State Agency
Bureau of Children's
  Services
Hoover Building
Des Moines, Iowa 50319

Court of Jurisdiction
District Court

Birth and Death Records
Division of Statistics
Department of Health
Des Moines, Iowa 50319

Marriage and Divorce
  Records
Same as Birth and Death

Archives/Records
Historical and Genealogical
  Library
Historical Building
East 12th and Grand Avenue
Des Moines, Iowa 50319

KANSAS

Information Department
State House
10th and Harrison Streets
Topeka, Kansas 66612

State Agency
Department of Social Services
Division of Children
2700 West 6th Street
Topeka, Kansas 66606

Court of Jurisdiction
District Court

Birth and Death Records
Bureau of Registration
  Statistics
6700 South Topeka Avenue
Topeka, Kansas 66620

Marriage and Divorce Records
Same as Birth and Death

Archives/Records
Historical Society Library
Memorial Building
Topeka, Kansas 66603

KENTUCKY

Information Department
State Capitol
Frankfort, Kentucky 40601

State Agency
Bureau for Social Services
275 East Main Street
6th Floor West
Frankfort, Kentucky 40601

Court of Jurisdiction
Circuit Court

Birth and Death Records
Office of Vital Statistics
Department of Health
275 East Main Street
Frankfort, Kentucky 40601

Marriage and Divorce Records
Same as Birth and Death

Archives/Records
State Library Archives
851 East Main Street
Frankfort, Kentucky 40601


LOUISIANA

Information Department
State Capitol
900 Riverside North
Baton Rouge, Louisiana 70804

State Agency
Office of Human Development
Adoption Program
333 Laurel, Room 704
Baton Rouge, Louisiana 70801

Court of Jurisdiction
District Court

Birth and Death Records
Division of Vital Records
P.O. Box 60603
New Orleans, Louisiana 70160

Marriage and Divorce Records
Same as Birth and Death


MAINE

Information Department
State House
Augusta, Maine 04333

State Agency
Department of Human Services
221 State Street
Augusta, Maine 04333

Court of Jurisdiction
County Probate Court

Birth and Death Records
Office of Vital Records
221 State Street
Augusta, Maine 04333

Marriage and Divorce Records
Same as Birth and Death

Archives/Records
State Library
L M A Building, State House
Augusta, Maine 04333


MARYLAND

Information Department
State House
State Circle
Annapolis, Maryland 21404
301-267-0100

State Agency
Social Services
    Administration
11 South Street
Baltimore, Maryland 21202

Court of Jurisdiction
Chancery Court

Birth and Death Records
Division of Vital Records
State Office Building
201 West Preston Street
P.O. Box 13146
Baltimore, Maryland 21203

Marriage and Divorce
    Records
Same as Birth and Death

Archives/Records
State Library
361 Rose Boulevard
Annapolis, Maryland 21401

MASSACHUSETTS

Information Department
State House
Beacon Street
Boston, Massachusetts 02133

State Agency
Department of Social Services
150 Causeway Street
Boston, Massachusetts 02114

Court of Jurisdiction
Probate Court

Birth and Death Records
Registrar of Vital Statistics
Room 103, McCormack Building
1 Ashburton Place
Boston, Massachusetts 02108

Marriage and Divorce Records
Same as Birth and Death

Archives/Records
State Library
341 State House
Boston, Massachusetts 02133

MICHIGAN

Information Department
Capitol Building
Lansing, Michigan 48933

State Agency
Department of Social Services
300 South Capitol Avenue
P.O. Box 3007
Lansing, Michigan 48909

Court of Jurisdiction
Probate Court

Birth and Death Records
Office of Vital Statistics
3500 North Logan Street
Lansing, Michigan 48914

Marriage and Divorce Records
Same as Birth and Death

Archives/Records
State Library
735 East Michigan Avenue
Lansing, Michigan 48933

MINNESOTA

Information Department
State Capitol
Aurora Avenue and Park Street
St. Paul, Minnesota 55155

State Agency
Department of Public Welfare
Centennial Office Building
St. Paul, Minnesota 55155

Court of Jurisdiction
County Family Court or Juvenile
    Division of District Court

Birth and Death Records
Department of Health/Vital
    Statistics
717 Delaware Street SE
Minneapolis, Minnesota 55440

Marriage and Divorce Records
Same as Birth and Death

Archives/Records
Historical Society Library
690 Cedar Street
St. Paul, Minnesota 55101

MISSISSIPPI

Information Department
New Capitol Building
Jackson, Mississippi 39205

State Agency
Department of Public Welfare
515 East Amite Street
P.O. Box 352
Jackson, Mississippi 39205

Court of Jurisdiction
Chancery Court

Birth and Death Records
Vital Records Unit
Board of Health
P.O. Box 1700
Jackson, Mississippi 39205

Marriage and Divorce Records
Same as Birth and Death

Archives/Records
Archives History Building
Capitol Green
Jackson, Mississippi 39205

MISSOURI

Information Department
State Capitol
Jefferson City, Missouri 65101

State Agency
Department of Social Services
P.O. Box 88
Jefferson City, Missouri 65103

Court of Jurisdiction
Circuit Court

Birth and Death Records
Bureau of Vital Records
Public Health and Welfare
Jefferson City, Missouri 65101

Marriage and Divorce Records
Same as Birth and Death

Archives/Records
State Library
308 East High Street
P.O. Box 387
Jefferson City, Missouri 65101

MONTANA

Information Department
Capitol Building
Helena, Montana 59601

State Agency
Social Services Bureau
111 Sanders
Helena, Montana 59601

Court of Jurisdiction
District Court/Tribal Court

Birth and Death Records
Bureau of Records and Statistics
Department of Health
Helena, Montana 59601

Marriage and Divorce Records
Same as Birth and Death

Archives/Records
State Library
930 East Lyndale Avenue
Helena, Montana 59601

NEBRASKA

Information Department
State Capitol
1445 K Street
Lincoln, Nebraska 68509

State Agency
Division of Social Services
P.O. Box 95026
Lincoln, Nebraska 68509

Court of Jurisdiction
County Court

Birth and Death Records
Bureau of Vital Statistics
301 Centennial Mall South
P.O. Box 95007
Lincoln, Nebraska 68509

Marriage and Divorce
    Records
Same as Birth and Death

Archives/Records
Historical Society Library
1500 and R Streets
Lincoln, Nebraska 68503

NEVADA

Information Department
State Capitol
Carson City, Nevada 89710
702-885-5000

State Agency
Welfare Division
251 Jeanell Drive
Capitol Mall Complex
Carson City, Nevada 89710

Court of Jurisdiction
District Court

Birth and Death Records
Health-Vital Statistics
Capitol Complex
Carson City, Nevada 89710

Marriage and Divorce Records
Same as Birth and Death

Archives/Records
State Library
401 North Caron Street
Carson City, Nevada 89711

NEW HAMPSHIRE

Information Department
State House
107 North Main Street
Concord, New Hampshire 03301
603-271-1110

State Agency
Bureau of Child/Family
    Services
Health and Welfare
Hazen Drive
Concord, New Hampshire 03301

Court of Jurisdiction
Probate Court

Birth and Death Records
Bureau of Vital Records
Health and Welfare
Hazen Drive
Concord, New Hampshire 03301

Marriage and Divorce
    Records
Same as Birth and Death

Archives/Records
State Library
20 Park Street
Concord, New Hampshire 03301

NEW JERSEY

Information Department
State House
Trenton, New Jersey 08625
609-292-2121

State Agency
Youth and Family Services
P.O. Box 510
Trenton, New Jersey 08625

Court of Jurisdiction
Superior or County Court

Birth and Death Records
Department of Health
Bureau of Vital Statistics
P.O. Box 1540
Trenton, New Jersey 08625

Marriage and Divorce Records
Same as Birth and Death

Archives/Records
State Library
185 West State Street
Trenton, New Jersey 08625

NEW MEXICO

Information Department
State Capitol
Santa Fe, New Mexico 87501

State Agency
Adoption Services
P.O. Box 2348
Santa Fe, New Mexico 87503

Court of Jurisdiction
District Court

Birth and Death Records
Vital Statistics Bureau
P.O. Box 968
Santa Fe, New Mexico 87503

Marriage Records
County Clerk where marriage
    performed

Divorce Records
County Clerk of District Court
    where granted

Archives/Records
State Library
300 Don Gaspar
Santa Fe, New Mexico 87501

NEW YORK

Information Department
State Capitol
Albany, New York 12234

State Agency
Department of Social Services
40 North Pearl Street
Albany, New York 12234

Court of Jurisdiction
Judge or Surrogate of court where
    order was made

Birth and Death Records
    (except New York City)
Bureau of Vital Records
Empire State Plaza
Tower Building
Albany, New York 12237

All New York City Boroughs
Bureau of Records
Department of Health
125 Worth Street
New York, New York 10013

Marriage and Divorce Records
(Addresses apply as indicated
above)

NORTH CAROLINA

Information Department
State Capitol
Raleigh, North Carolina 27611

State Agency
Children's Services Branch
Department of Human Resources
325 North Salisbury Street
Raleigh, North Carolina 27611

Court of Jurisdiction
Clerk of Superior Court

Birth and Death Records
Department of Human Services
Vital Records Branch
P.O. Box 2091
Raleigh, North Carolina 27602

Marriage and Divorce Records
Same as Birth and Death

Archives/Records
State Library
109 East Jones Street
Raleigh, North Carolina 27611

NORTH DAKOTA

Information Department
State Capitol
Bismarck, North Dakota 58505
701-224-2000

State Agency
Children and Family Services
Russel Building
Box 7
Bismarck, North Dakota 58505

Court of Jurisdiction
District Court

Birth and Death Records
Vital Records
Department of Health
Bismarck, North Dakota 58505

Marriage and Divorce Records
Same as Birth and Death

Archives/Records
State Library
Highway 83, North
Bismarck, North Dakota 58505

OHIO

Information Department
State House
Broad and High Streets
Columbus, Ohio 43215
614-466-2000

State Agency
Department of Welfare
Bureau of Children's Services
30 East Broad Street, 30th Floor
Columbus, Ohio 43215

Court of Jurisdiction
Probate Court

Birth and Death Records
Division of Vital Statistics
G-20 Ohio Departments Building
65 South Front Street
Columbus, Ohio 43215

Marriage and Divorce Records
Same as Birth and Death

Archives/Records
State Library
65 South Front Street
Columbus, Ohio 43215


OKLAHOMA

Information Department
State Capitol
2302 Lincoln Boulevard
Oklahoma City, Oklahoma 73105
405-521-2011

State Agency
Division of Child Welfare
P.O. Box 25352
Oklahoma City, Oklahoma 73125

Court of Jurisdiction
District Court

Birth and Death Records
Vital Statistics Section
Northeast 10th Street and
    Stonewall
P.O. Box 53551
Oklahoma City, Oklahoma 73105

Marriage and Divorce Records
County where license was issued,
    or divorce granted

Archives/Records
Oklahoma Historical Society
Historical Building
2100 North Lincoln Boulevard
Oklahoma City, Oklahoma 73105

OREGON

Information Department
State Capitol
Salem, Oregon 97310

State Agency
Adoption Services
Department of Human Resources
198 Commercial Street SE
Salem, Oregon 97310

Court of Jurisdiction
Circuit Court

Birth and Death Records
Vital Statistics Section
Health Division
P.O. Box 116
Portland, Oregon 97207

Marriage and Divorce Records
Same as Birth and Death

Archives/Records
State Library
Summer and Court Streets
Salem, Oregon 97310


PENNSYLVANIA

Information Department
Main Capitol Building
Harrisburg, Pennsylvania 17120

State Agency
Public Welfare
Office of Children and Families
P.O. Box 2675
Harrisburg, Pennsylvania 17120

Court of Jurisdiction
Court of Common Pleas

Birth and Death Records
Division of Vital Statistics
Central Building
101 South Mercer Street
P.O. Box 1528
Newcastle, Pennsylvania 16103

Marriage and Divorce Records
Same as Birth and Death

Archives/Records
State Library
Walnut Street and Common-
   wealth Avenue
Harrisburg, Pennsylvania 17120

RHODE ISLAND

Information Department
State House
82 Smith Street
Providence, Rhode Island 02903

State Agency
Children and Families
   Department
610 Mt. Pleasant Avenue
Providence, Rhode Island 02908

Court of Jurisdiction
Adoptees over 18: Probate
Adoptees under 18: Family

Birth and Death Records
Division of Vital Statistics
Department of Health
Cannon Building
75 Davis Street
Providence, Rhode Island 02908

Marriage and Divorce Records
Same as Birth and Death

Archives/Records
State Library
State House
82 Smith Street
Providence, Rhode Island 02903

SOUTH CAROLINA

Information Department
State House
Columbia, South Carolina 29211

State Agency
The Children's Bureau
800 Dutch Square Boulevard
Building D
Columbia, South Carolina 29211

Court of Jurisdiction
Family Court

Birth and Death Records
Division of Vital Records
Department of Health
2600 Bull Street
Columbia, South Carolina 29201

Marriage and Divorce Records
Same as Birth and Death

Archives/Records
State Library
1500 Senate Street
Columbia, South Carolina 29201

SOUTH DAKOTA

Information Department
Capitol Building
Pierre, South Dakota 57501

State Agency
Department of Social Services
Richard K. Kneip Building
Pierre, South Dakota 57591

Court of Jurisdiction
Circuit Court

Birth and Death Records
Health Department
Joe Foss Office Building
Pierre, South Dakota 57501

Marriage and Divorce Records
Same as Birth and Death

Archives/Records
State Library
Soldiers and Sailors Memorial
  Building
East Capitol Avenue
Pierre, South Dakota 57501

TENNESSEE

Information Department
State Capitol
Nashville, Tennessee 37219

State Agency
Department of Human Services
111-19 7th Avenue, North
Nashville, Tennessee 37203

Court of Jurisdiction
Adoptions prior to 1950:
Probate or County Courts
Adoptions after 1950:
Chancery or Circuit Courts

Birth and Death Records
Division of Vital Statistics
Cordell Hull Building
Nashville, Tennessee 37219

Marriage and Divorce Records
Same as Birth and Death

Archives/Records
State Library
403 7th Avenue, North
Nashville, Tennessee 37219

TEXAS

Information Department
State Capitol
Austin, Texas 78711

State Agency
Department of Human Resources
706 Bannister Lane
P.O. Box 2960
Austin, Texas 78769

Court of Jurisdiction
District Court

Birth and Death Records
Bureau of Vital Statistics
Department of Health
1100 West 49th Street
Austin, Texas 78701

Marriage and Divorce Records
Same as Birth and Death

Archives/Records
State Library
1201 Brazos
Austin, Texas 78701

UTAH

Information Department
State Capitol
Salt Lake City, Utah 84114

State Agency
Department of Family Services
150 West North Temple
Salt Lake City, Utah 84103

Court of Jurisdiction
District Court

Birth and Death Records
Bureau of Vital Statistics
150 West North Temple
P.O. Box 150
Salt Lake City, Utah 84103

Marriage and Divorce Records
County Clerk where license was
   issued and decree granted

Archives/Records
Historical Society Library
603 East South Temple
Salt Lake City, Utah 84102

VERMONT

Information Department
State House
State Street
Montpelier, Vermont 05602

State Agency
Agency of Human Services
103 South Main Street
Waterbury, Vermont 05676

Court of Jurisdiction
District Probate Court

Birth and Death Records
Public Health Statistics
   Division
115 Colchester Avenue
Burlington, Vermont 05401

Marriage and Divorce
   Records
Same as Birth and Death

Archives/Records
Historical Society Library
Pavillion Office Building
109 State Street
Montpelier, Vermont 05602

VIRGINIA

Information Department
State Capitol
Richmond Square
Richmond, Virginia 23219

State Agency
Division of Social Services/
   Welfare
8007 Discovery Drive
Richmond, Virginia 23288

Court of Jurisdiction
Circuit Court

Birth and Death Records
Vital Records and Health
   Statistics
James Madison Building
P.O. Box 1000
Richmond, Virginia 23208

Marriage and Divorce Records
Same as Birth and Death

Archives/Records
State Library
12th and Capitol Streets
Richmond, Virginia 23219

WASHINGTON

Information Department
State Capitol
Olympia, Washington 98504

State Agency
Department of Health and
   Children's Services
Office Building #2
Olympia, Washington 98504

Court of Jurisdiction
Superior Court

Birth and Death Records
Vital Records LB-11
P.O. Box 9709
Olympia, Washington 98504

Marriage and Divorce Records
Same as Birth and Death

Archives/Records
State Library
Archives Division
Olympia, Washington 98504

WEST VIRGINIA

Information Department
State Capitol
1800 Kanawha Boulevard East
Charleston, West Virginia 25305

State Agency
Department of Welfare
1900 Washington Street East
Charleston, West Virginia 25305

Court of Jurisdiction
Circuit Court

Birth and Death Records
Division of Vital Statistics
State Office Building No. 3
Charleston, West Virginia 25305

Marriage and Divorce Records
Same as Birth and Death

Archives/Records
Historical Society
Cultural Center
Capitol Complex
Charleston, West Virginia 25305

WISCONSIN

Information Department
State Capitol
Capitol Square
Madison, Wisconsin 53702

State Agency
Department of Health and Social
   Services
1 West Wilson Street
Madison, Wisconsin 53702

Court of Jurisdiction
Circuit Court

Birth and Death Records
Bureau of Health Statistics
P.O. Box 309
Madison, Wisconsin 53701

Marriage and Divorce Records
Same as Birth and Death

Archives/Records
Historical Society
University of Wisconsin
816 State Street
Madison, Wisconsin 53706

WYOMING

Information Department
State Capitol
Capitol Avenue at 24th Street
Cheyenne, Wyoming 82001

State Agency
Division of Public Assistance
Hathaway Building
Cheyenne, Wyoming 82002

Court of Jurisdiction
District Court

Birth and Death Records
Vital Records Service
Hathaway Building
Cheyenne, Wyoming 82002

Marriage and Divorce Records
Same as Birth and Death

Archives/Records
State Library
Supreme Court and Library
   Building
Cheyenne, Wyoming 82001

# INDEX

Abduction of children, 86
Accepting of results if you
    succeed, 10, 107
Adler, Allan, 51
Adoptee/birth parent search,
    105–21
  beginning notes, 107–109
  the need to know, 106
  petitioning the court, 120–21
  reunion registries, 112, 116–19
  state contacts for adoption
    records, 106, 221–39
  support for, 109–15
  tracing birth physicians, 120
Adoptees-in-Search (AIS), 112
Adoptee's Liberty Movement
    Association (ALMA),
    105–106, 112–13, 119
Adoption:
  birth certificates, 24–27, 120
  confidentiality of records, 34,
    105–106, 120
  search for birth parent, *see*
    Adoptee/birth parent
    search
  state contacts for adoption
    records, 221–39
*Adoption Fact Book, The*, 111

Adoption Search Institute
    (ASI), 113–14
After-school programs, 98
Agencies, public and private,
    assistance from, 14–17, 41
Air Force, U.S., 30, 55
Airplane flight, births on
    international, 27–28
Airplane ownership records,
    32, 60
Aka (also known as), 12
Alcohol, Tobacco, and Firearms
    (ATF, part of Department
    of the Treasury), 58
Alien children adopted by U.S.
    citizens, birth records of,
    26–27
ALMA International Reunion
    Registry, 113, 117
American Association of
    Retired Persons (AARP), 65
American Civil Liberties Union
    (ACLU), 51
American Express, 79
Armed forces:
  locator services, 55–56
  records of, 30, 56
  Selective Service, 59

Army, U.S., 30, 55
Associations, 65–66
Assumed name, 12
Automobiles, *see* Cars
*Ayer's Directory of Newspapers and Periodicals,* 64, 67

Bacon's clipping service, 72
Bankruptcy court, 55
Birth physicians 120
Birth records, 24–28, 128–29
    of adoptees, 24–25, 26–27, 105–106, 120
    of alien children adopted by U.S. citizens, 26–27
    of births on international territory, 27–28
    facts to provide when requesting, 25
    of foreign countries, 28
    state addresses to address requests to, 203–220
    of U.S. citizens born in foreign countries, 25–26, 27
Birth parents search, *see* Adoptee/birth parent search
Block Parent safety program, 98
Block patrols, 99–100
Boards of licensing, state, 47
Boat licenses, 47
Book sources, 72–73
Brooklyn Business Library, 76
Bureau of the Census, 58
Burrelle's Press clipping service, 72
Business credit reporting agencies, 80

*Business Guide to Corporate Executives, The,* 70
Business names, fictitious, 32

California Child, Youth, Family Coalition, 89
Carroll, John M., 73
Cars:
    accident reports, 46
    registration records, 32
    safety tips for adults, 101–103
    safety tips for children, 92–93
    state records, 44, 198–202
Catalog mailing lists, 65
Census Bureau, *see* Bureau of the Census
Certified mail, 53–54
Chambers of Commerce, 72
Charles, Jan, 84
Charles, Peggy, 84
Chicago, Illinois, 87
    Police Department, 83
Child Find, 87
Child Search, 89
Child support:
    District Attorney's Family Support Unit, 32–33
    Federal Office of Child Support Enforcement, 59–60
Children:
    abduction of, 86
    after-school programs, 98
    block patrols for, 99–100
    fingerprinting of, 100
    kidnapping of, 86–87
    missing, 21, 57–58, 83–90, 134–36

neighborhood safety, 98
organizations concerned
    with missing and abused,
    addresses of, 134–36
overseas, 101
parent alerts, 100
physical abuse of, 96–97
"play it safe" rules for, 94
safe homes for, 98
safety tips for, 91–101
school crime prevention
    curriculum for, 99
sexual molestation of, 95–96
sources of information on
    child safety, 100–101
teaching safety rules to, 91–92
teaching them how to
    handle strangers, 92–93, 95
Children's Rights of
    Pennsylvania, 89
Church of Latter-day Saints, see
    Family History Library of
    the Church of Jesus Christ
    of Latter-day Saints
Citizenship, U.S., 26
City jail records, 60–61
City records, 20–22, 61
Civil court cases, 35, 36
Clipping services, 72
Coast Guard, U.S., 28, 30, 56
College libraries, 21
Computers, 74
Concerned United
    Birthparents (CUB), 114,
    119
    reunion registry of, 117
Confidential Information Sources,
    Public and Private (book), 73
Consular Report of Birth,
    25–26
Consumer credit reporting
    agencies, 78–79
Corporate records, 45

County Public Administrator's
    Office (CPAO), 29
County records, 23–41
    birth records, 24–28
    civil and criminal court
        records, 33–38
    death records, 28–30
    District Attorney's Family
        Support Unit, 32–33
    fictitious business names,
        32
    Grantor/Grantee records,
        33
    licenses and permits, 22, 41
    marriage records, 30–31
    prison records, 60–61
    public welfare records,
        39–41, 60
    real property records,
        31–32
    sheriff's office records, 23–24
    unsecured property records,
        32
    voter registration records, 39
    workmen's compensation
        records, 41
Court records, 33–37, 54–55
Courts, petitioning, regarding
    adoption records, 120–21
Credit card company
    investigative and fraud
    divisions, 79
Credit reporting agencies,
    78–79
Crime prevention curriculum,
    99
Criminal records, 21–22, 23–24,
    56–58
    of courts, 33–38
    prison records, 60–61
Criss-Cross directory, see
    Haines Criss-Cross
    directory

Custodial parents, tips for, regarding kidnapping, 88–89
Custody decree, 88, 89

Dahmer, Jeffery, 83
Damages, civil suit, 35
DataSearch, 77
Date of birth (DOB), 12–13, 42
Day-care centers, 95
Deadlines, setting, 5
Death records, 28–30, 128–29
    county records, 29
    requests for, data to include in, 29–30
    Social Security Administration, 28
    state addresses to address requests for, 203–220
    of U.S. citizens who die in foreign countries, 30
Department of Health and Human Services, U.S., Federal Office of Child Support Enforcement, 59–60
Department of Motor Vehicles, 32
Department of State, U.S., 26, 28, 30, 101
Deserted areas, staying away from, 94
Devious tactics, 11–12
Directories, searching, 17
District attorney's office records, 23–24, 60
    Family Support Unit, 32–33
District Court, U.S., 54
Divorce Court records, 32–33, 128–29
    state addresses to address requests for, 203–220
Doctors, birth, 120

Driver's license, 102
Driver's License Bureau, 36, 43
Driver's license records, 42–44, 77
    addresses of state agencies handling, 198–202
Dun & Bradstreet, 80

Educational channels, 62–64
Emergency telephone numbers, 97, 103–104
Expenses:
    fees for, *see specific types of records*
    of "friendly" search, 6–7
    records of, 7
Express mail, 54

Family History Library Catalog, 123, 126, 143
Family History Library of the Church of Jesus Christ of Latter-day Saints, 76, 122–27
    Ancestral File, 125
    branch libraries, 143–84
    Catalog of, 123, 126, 143
    copy services, 124
    Family Registry, 125
    film circulation, 124
    International Genealogical Index, 123, 126, 143
    library classes, 125
    little-used files, 124
    Personal Ancestral File, 125
    preparation for using, 126
    professional searchers of, 126, 185
    publications of, 124
    rules of, 127
    sharing information with, 126

supplies, 126
working schedules of,
    127
Federal Aviation Authority
    (FAA), 32, 60
Federal Bureau of
    Investigation (FBI), 71
missing persons and, 57, 85,
    87, 88
Federal help, *see* National
    Archives
Federal Office of Child
    Support Enforcement
    (FOCSE), 59–60
Federal records, 49–61
    Alcohol, Tobacco, and
        Firearms, 58
    bankruptcy court, 55
    Federal Aviation Authority,
        32, 60
    Federal Office of Child
        Support Enforcement,
        59–60
Freedom of Information and
    Privacy Acts (FOIA) and,
    22, 49–51
    Government Printing Office,
        58–59
    Internal Revenue Service, 58
    Interstate Commerce
        Commission, 58
    military locators, 55–56
    military records, 30, 56
    National Crime Information
        Center, 21–22,
        57–58
    Postal Service, 53–54
    prison records, 61
    Selective Service, 59
    Social Security
        Administration, 28, 51–53
    U.S. District Court, 54
    U.S. Marshal's office, 56–57

Felony, 35, 88
Fictitious business names, 32
Fingerprinting children, 100
Fire department, 104
Fish and Game commissions,
    state, 47
Fisher, Florence, 105
Fishing licenses, 47
Florida Department of Law
    Enforcement, 87
Foreign country(ies):
    birth record of person born
        in, who is U.S. citizen at
        birth, 25–26
    birth records in, 28
    death records of U.S. citizens
        who die in, 30
Freedom of Information Act
    (FOIA), 22, 49–51
"Friendly" search, 6–11
Fugitives from the law, 56–57
FYI-Telephoning free, 67, 74

Gacy, John Wayne, 83
Genealogists, 126, 185–97
General Delivery mail, 54
General Services
    Administration, 56
Goodman, Ellen, 87
Government Printing Office,
    59, 91, 104
Grand jury, records of, 37–38
Grantor/Grantee records, 33

Haines Criss-Cross directory,
    17, 18, 63
Halloween and safety, 94
*Handbook for Single Adoptive
    Parents, The*, 110
Helping Hand program, 98
Highway patrol, state, 46
Hitchhiking, 94
Hot lines, 89–90, 96–97

Immigration and
  Naturalization Act, 27
Immigration and
  Naturalization Service, 27
Independent Search
  Consultants, 113
Indictment, grand jury, 37–38
*Information Book, The*, 59
Information on Demand, 77
Information sources, 75–79
Infosearch, Inc., 78
Inquiry kit, 8, 16, 21
Insurance records, 66
Internal Revenue Service (IRS),
  58
International Genealogical
  Index, 123, 126, 143
International Soundex Reunion
  Registry (ISRR), 117–18
International territory, birth
  records for births on,
  27–28
Internet, 74, 75
Interstate Commerce
  Commission (ICC), 58
Investment ventures, persons
  involved in, 77

Johnson, Lyndon, 49
Journal of the search, 6, 9
Justice Department, U.S., 86, 89

Kammandale Library, 119
Kansas City Adult Adoptee
  Organization (KCAAO),
  114
Kidnapping, 86–87
Lace, clipping service, 72
Law-enforcement agencies,
  local, 103
  child safety and, 96–101
Law libraries, 21

Lawyers, assistance of, 88, 89,
  120–21
Letter of inquiry, 15–16
  Freedom of Information
    Act/Privacy Act, 50
  to Public Services/Welfare
    Department, 40
  to Social Security
    Administration, 52
  to voter registration office,
    38, 39
Liberal Education for Adoptive
  Families, 119
Librarians, reference, 9, 18–19,
  20, 66, 69
Library(ies):
  cardholder names, 20–21
  genealogical, 122–27, 143–84
  joining the, 9, 18–19
  telephone directories, 18
Licenses:
  city and county, 22, 41
  driver's records, 42–44, 77,
    198–202
  federal, 58
  state, 42–44, 47, 198–202
*Litigation Under the Federal
  Freedom of Information Act
  and Privacy Act*, 51
Local Court and County
  Record Retrievers, 77
Locking doors and windows,
  102
Los Angeles telephone
  directories, Greater, 18

McGruff Safe Home program,
  98
Magazine subscription lists,
  66–67
Mail, 53–54
Mail-order catalogs, 65
Mailing supplies, 6, 9

Maine Reunion Registry, 118
Marine Corps, U.S., 56
Marriage records, 30–31,
    128–29
  state addresses to address
    requests for, 203–220
Marshal, U.S., 56–57
MasterCard, 80
Medical associations, 120,
    137–39
Michigan Reunion Registry, 118
Minnesota Reunion Registry,
    119
Missing Children Act of 1982,
    88
Missing Children's Help
    Center, 89
Missing persons, 21, 57–58,
    81–90
  children, 21, 57–58, 81–90,
    134–36
  name changers, 82–83
  National Crime Information
    Center and, 21, 57–58, 84–85
  reports of, 83–84
Moody's, 80
Mormons, library of, *see*
    Family History Library of
    the Church of Jesus Christ
    of Latter-day Saints
Motor vehicles, 77
  driver's license records,
    42–44, 77, 198–202
  registration records, 32
  state records, 44, 198–202
  *see also* Cars
Municipal Court records, 35–36
Name of subject, 12, 42–43
  common names, 16–17
  fictitious business names, 32
  name changers, 82–83
*Names and Numbers* (book), 73

National Adoption Support
    Groups, 110–12
National Archives, 76
National Center for Health
    Statistics, 128
National Center for Missing
    and Exploited Children
    (NCMEC), 85–90
*National Clearinghouse
    Directory, The*, 111
National Crime Information
    Center (NCIC), 21–22, 57,
    84–85, 88
National Data Research Center
    76
*National Directory of Addresses
    and Telephone Numbers*, 67,
    72
National Driver's Registration
    Service, 45
National Hot Line, 89, 90
Navy, U.S., 30, 56
Neighborhood safety, 98–100
Neighborhood Watch, 100
Neighbors, contacting former,
    16
Nevada Reunion Registry,
    119
New Jersey Reunion Registry,
    119
New Jersey State Missing
    Persons Commission,
    84
New York Public Library,
    127
Newspapers:
  clipping services, 72
  directories, 64
  morgues, 64–65
  personals, 64
Nolan, Danny, story of, 1–4
Norland, Rod, 73
North Carolina, 100

Orphan Voyage, 115
Orphan Voyage Reunion
    Registry, 118

Parent(s):
    alerting, of child absent from
        school, 100
    custodial, tips for, regarding
        kidnapping, 88–89
    fingerprinting children, 100
Parental kidnapping, 86–87, 88
Paton, Jean, 115
Pay telephones, 71, 72
Periodical subscription lists,
    66–67
Permits, 22, 41, 47
Personals, newspaper, 64
Petitioning courts for access to
    adoption records, 120–21
Physical abuse of children,
    96–97
Pilots, 60
Place of birth (POB), 13
Police departments, 21–22, 46,
    97, 103–104
Political leanings, 4
Post office boxes, 54
Postal Service, U.S., 53–54
Postcards, 6, 7, 9, 15–17
Prison records, 60–61
Privacy Act of 1974, 50–51
Private investigators, xiv,
    11–12, 71
Probate records, 36
Professional associations,
    65–66
Professional licenses, 47
Profile of your subject:
    creating a, 7–8, 107
    items to include, 9
    sending, to agencies with
        letter of inquiry, 15, 39–41,
        55–56

Property records, 31–32, 76–77
Prothonotary, 33
Psychological counseling, 89
Public defender's office,
    records of, 24
Public information retrieval
    service, 77
Public Record Research
    Library, 77

R. R. Bowker's National
    Directory of Weekly
    Newspapers, 64
Real property records, 31–32,
    76–77
Recordkeeping for your search,
    6, 7, 9, 10
Red Book, The, 70
Registered mail, 54
Registries of persons with
    outstanding achievements,
    69–70
Religious affiliations, 4, 69
Reunion registries for
    adoptee/birth parent
    search, 112–13, 116–21
Riller, Mary Jo, 115
Ruffino, Richard, 84–85

Safe homes for children, 98
Safety tips, 91–104
    for adults, 101–104
    for children, 91–101
    government pamphlets on,
        104
Sales tax board, state, 47–48
Salvation Army, xiv–xv, 68–69
Sanders, Pat, 113
Scenario for adoptee/birth
    parent search, 107–108
Schools:
    after-school programs, 98

crime prevention
curriculum, 99
security measures of, 93
Secretary of Defense, 30
Securities and Exchange
Commission (SEC), 45
Selective Service, 59
Sexual abuse of children,
95–96
addresses of organizations
dealing with, 134–36
Sheriff's office records, 23–24
Small Claims Court records, 36
Social Security Administration,
28, 51–53
Social Security numbers, 13$n$,
52–53, 132–33
Society for Youth Victims, 89
Standard & Poors, 80
State Franchise Tax Board, 46
State records, 42–48
corporate, 45
driver's license records,
42–44, 77, 198–202
motor vehicle, 44, 198–202
permits and licenses, 47
prison records, 60–61
state police/highway patrol,
46
state sales tax board,
47–48
state tax boards, 46
Strangers:
safety tips for adults, 101,
102, 103
teaching children how to act
toward, 92–93, 95
Subpoena duces tecum (court
order), 34, 54
Subscription lists, magazine,
66–67
Superintendent of Documents,
59, 104

Superior Court records, 35
Support groups, 7, 88, 109–15
Supreme Court, U.S., 54

Take a Bite Out of Crime—How
to Protect Your Children, 91
Task Force on Parental
Abduction, 89
Tax assessor's office, property
records of, 31–32
Tax board, state, 46
sales tax board, 47–48
Tax number, federal, 58
Telephone calls:
collect, 41
secure, 71–72
time zones and cost of, 6
Telephone company, 70–71
Telephone directories, 18
Haines Criss-Cross, 17, 18,
63
national, 72
Telephone numbers:
emergency, 97, 103–104
hot lines, 89–90
Toll-Free (800), 67
unlisted, 70–71
Telephone taps, 71
Telephone traps, 71–72
Tracers of America, 81
Traffic Court records, 36
Travel, safety tips for,
102–103
Triadoption Library, 115
TRW Consumer Relations, 78
Tufts University, 100

Unions, 65
U.S. Marshal, 56–57
University libraries, 21
Unlawful Flight to Avoid
Prosecution (UFAP)
warrant, 88

Unsecured property records, 31–32
Utilities, 70

Visa International, 80
Vital information, 12–13
Vital records, writing for, 128–29
    addresses by state, 203–220
    *see also* Birth records; Death records; Divorce Court records; Marriage records

Voter registration records, 38–39

Welfare department records, 39–40, 60
*Who's Who in America*, 69, 70
Wills, probate records, 36
Witnesses, government-protected, 34
Women, safety tips for, 101, 103
Workmen's compensation records, 41